Lecture Notes in Control and Information Sciences

Edited by M. Thoma and A. Wyner

For information about Vols. 1-96 please contact your bookseller or Springer-Verlag

Lecture Notes in
Control and
Information Sciences
Edited by M. Thoma and A. Wyner

169

V. M. Kuntzevich, M. Lychak

Guaranteed Estimates, Adaptation and Robustness in Control Systems

Springer-Verlag
Berlin Heidelberg GmbH

Authors
V. M. Kuntzevich
M. M. Lychak

Ukrainian Academy of Sciences
V. M. Glushkov Institute of Cybernetics
P. O. 252207
40 Prospect Academika Gluchkova
Kiev
CIS

ISBN 978-3-540-54925-3 ISBN 978-3-540-46598-0 (eBook)
DOI 10.1007/978-3-540-46598-0

Originally published by Springer-Verlag Berlin Heidelberg New York in 1992

Typesetting: Camera ready by authors

61/3020-5 4 3 2 1 0 Printed on acid-free paper.

CONTENTS

INTRODUCTION

The majority of real control plants feature considerable uncertainty regarding both the properties (parameters) of these plants and disturbances affecting them. In many cases, the a priori estimates of parameters are rather rough thus preventing the obtaining of practically reasonable solutions of problems of analysis and synthesis of respective control systems. This circumstance leads to the need to use one or other parametric identification procedure which can be realized either at the preliminary identification stage i.e. prior to the realization of the control process itself (what is obviously admissible only for the class of stable controlled plants), or within the framework of adaptive control systems in which the both processes: the studying (identification) and the control proper are carried out simultaneously. This approach is required specifically in solving the problem of control of unstable plants.

It might be well to point out that the majority of scientists studying control processes under uncertainty conditions use stochastic (probabilistic) uncertainty models until the present time. They do it not because the processes of uncertainty actually have a stochastic nature. In many cases this is the result of a certain inertia of thought, the action of an established stereotype.

The assumption of stochastic nature of uncertainty can not be accepted at least in two cases: 1) when the volume of a priori experimental data on the nature of the uncertain factors is so small that it does not allow the conclusion on the availability of stable statistical characteristics, 2) when it is known a priori that the uncertainty basically can not be considered to be produced by some probabilistic mechanism. As correctly noted by R.Kalman: "It would be a great untruth to claim that ... the whole uncertainty arises by virtue of the mechanism of statistical choice. The nature does not conform to the rules of traditional probability".

It is this reasoning that has given rise to the advent in the last decade of the method of obtaining guaranteed estimates of the vectors of parameters or (and) the state of the processes being controlled. The first chapter of the book calling the attention of the reader is devoted to the above range of problems.

Another important problem to which **Chapter 2** of the present book is dedicated is the problem of constructing adaptive control systems in which (in view of one or another reason) it is necessary to combine in time both the process of improvement of the accuracy of the estimates of the controlled plant parameters (and in some cases, of the guaranteed estimates of its state vector inaccessible for a direct measurement) and the process of the plant control proper. The simultaneous realization of these two processes has a significant mutual effect on the both processes.

The presence of uncertainty as regards the non-controllable disturbances (noise) acting upon the controlled plant results in incorrectness of the problems of analysis and synthesis of systems controlling such plants even given ideal estimates of their parameters. To eliminate this uncertainty, a gaim approach to the control synthesis problem statement is used i.e. the assumption of "ill-intentions" of the nature(of the environment generating these disturbances) is introduced. I.e., the nature tends (within the limits of specified restrictions) to maximize the performance index which the system designer seeks to minimize by selecting control. Thus, the problem of control synthesis is formulated as a minimax problem and some results in solving this problems with different statements of control synthesis problems are presented in **Chapter 2** .

The result in solving the identification problem under real conditions (irrespective of the identification algorithms used in this case and irrespective whether this solution is obtained within the framework of adaptive control systems or at a stage of preliminary (independent) identification) is always the obtaining of only some of parameter estimates. This takes place already even if the solution is sought for on a final time interval and in the presence of one or other disturbance (noise). In the event that these estimates are in some sense reasonably close to the true values of the parameters, it is allowable to identify these estimates with the true values of the parameters. When this takes place, no special problems arise, as a rule, with the analysis and synthesis of control systems for such plants. However, when the control system designer has no reasons to neglect the errors in solving the identification problems, then the situation is substantially complicated since the presence of such "residual" irremovable in principle uncertainty concerning the characteristics of the controlled plants is equivalent to the necessity to solve the problems of analysis and synthesis not for one specific plant but for a complete class of such plants, i.e. it leads us to the necessity to solve the problem of the robustness of control systems in general and of adaptive control systems in particular. Some problem statements and solutions of the control system robustness problem are set forth in the last third chapter.

This book is based on results of a research conducted by the authors at the V.M. Glushkov Institute of Cybernetics of the Ukrainian Academy of Sciences which have been rather widely discussed by the authors with their colleagues not only at the Institute. No doubt, the results of the discussions have been not only beneficial for the improvement of the style of their statement, but have had a positive impact on the direction of the studies themselves.

We should like gratefully acknowledge the support which was rendered to the authors by Professor M. Thoma during preparation of this book for publication.

Vsevolod M. Kuntsevich
Michail M. Lychak

Kiev, May, 1991.

CHAPTER 1

GUARANTEED ESTIMATES OF PARAMETER AND STATE VECTORS

"Throw off the whole of the impossible and then what remains is the truth".

Conan Doyle

1.1. Guaranteed Estimates of Linear Systems Parameters

Automatic control of any plant assumes that the control system designer has a more or less adequate mathematical model of the plant which is specified in many cases only up to a parameter vector. Therefore, the first problem which faces the control system designer is the problem of refinement of the a priori estimates of this parameter vector, i.e. the parameter identification problem. By virtue of a number of reasons mentioned already in the Introduction, we shall consider below only nonstochastic method of its solution which will prevent us from using stochastic methods for its solution described comprehensively in a number of papers which are now already classical (e.g., ref. to [1]-[7]). In the present chapter, the case will be considered when the parametric identification process can be carried out independently on the process of control proper and comes before to it in time. Thus, let us consider first the problem of determining the guaranteed estimates of the parameter vector for the simplest class of control plants, namely of linear plants without memory with the state described in discrete time by the equation

$$y_n = L^T U_n + f_n \quad , \qquad n = 1,2,\ldots, \tag{1}$$

where U_n is l-dimensional control vector ("input" of the plant), L is l -dimensional vector of constant but unknown parameters, f_n is scalar uncontrollable disturbance (noise), y_n is the measured scalar ("output"of the plant).

Let us assume that only a priori estimate

$$L \in \Omega_0 , \tag{2}$$

is known concerning parameter vector L, where Ω_0 is a given bounded convex set.

For disturbance f_n, given is also only its a priori estimate

$$f_n \in \mathcal{f} , \quad \forall\, n>0 , \tag{3}$$

where $\mathcal{f}$ is a preset bounded convex set determined in the following way:

$$\mathcal{f} = \{ f \mid |f| \leqslant \Delta = \mathrm{const} \} . \tag{4}$$

It is assumed that vector U_n is formed generally with due regard for constraints

$$U_n \in \Omega , \tag{5}$$

where Ω is a specified set in control space.

Such a priori information practically predetermines the use of the set identification procedure which enables a guaranteed estimate of parameter vector L to be obtained at each n-th step in form of its belonging to some set Ω_n.

Let us consider first the solution of the passive identification problem when input signals U_n are assumed to be known but they are generated in the way unknown to the investigator. Solution of this problem should be based both on the use of a priori information (equation (11) and estimates (2) and (3)) and on the use of a posteriori information (the results of measurements of values U_n and y_n).

Let us construct a recurrent procedure for the refinement of estimates of vector L. Assume that estimate L at the $(n+1)$-th step is known in the form:

$$L \in \Omega .\qquad (6)$$

with n=1 . The role of such estimate plays estimate (2).

Once the values U_{n+1} and y_{n+1} are measured, we obtain from (1) the estimate of vector L in the form:

$$L \in \tilde{\Omega}_{n+1} = \{ L |\ U^T_{n+1} L + f_{n+1} - y_{n+1} = 0 \}. \qquad (7)$$

From two consistent estimates (6) and (7), we obtain a new, generally improved, estimate Ω_{n+1} , which is described by the set evolution equation [8]-[11]:

$$L \in \Omega_{n+1} = \tilde{\Omega}_{n+1} \cap \Omega_n , \qquad n = 0,1,2,\ldots, . \qquad (8)$$

It should be emphasized here, that unlike the traditional identification problem solution methods which allow to obtain only approximate "point" estimates of parameters, estimate (6) is a guaranteed one in the sense that true values of parameters sought for are known to belong to a set Ω_{n+1} . When this takes place, all elements (points) of the set Ω_{n+1} are equal in rights since there are no preferences between the elements of the set.

An important property of the sequence of estimates (8) is that these estimates are non-deteriorating, i.e.

$$\Omega_{n+1} \subseteq \Omega_n \qquad (9)$$

even through non-informative changes are possible with some n when no improvements of the estimate take place. In the latter case $\Omega_n \subset \tilde{\Omega}_{n+1}$ and $\Omega_{n+1} = \Omega_n$. Clearly, the number of non-informative measurements grows with a sufficiently good accuracy and hence the obtained estimates may appear hereinafter to be non-improving with sufficiently large n . Only in some special cases, sets Ω_n with large n can contract (degenerate) to a single-point set containing only one point corresponding to the true value of $\overset{*}{L}$ of the plant parameter vector.

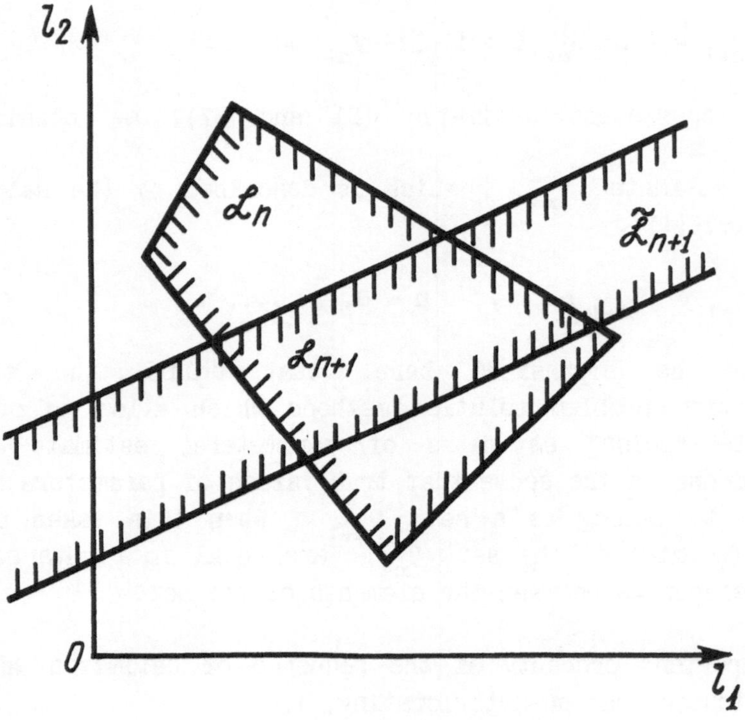

Fig. 1. Intersection of convex polyhedra

Another important property of the sequence of estimates (8) is as follows: when a priori estimate ϱ_0 is a convex polyhedron, then all following estimates ϱ_n are also convex polyhedra (ref. to **Fig. 1**) since the class of convex polyhedra is closed with respect to the intersection operation.

Hereinafter, we shall assume that ϱ_0 is a convex polyhedron without stipulating this each time separately.

Let us now dwell on the definition of the necessary and sufficient conditions with fulfillment of which the sequence of estimates ϱ_n degenerates into a single-point set. We consider first the degenerate case in which set f is empty and which is realized with $\Delta = 0$.

Then we obtain from (1) at $\Delta = 0$ that this equation describes some hypersurface in space $\{L\}$ for each n. Let us note that when directing vectors U_n of each of the hypersurfaces are linearly-independent on each other then , as a result of the execution of each of the set intersection operations, we obtain set ϱ_{n+1} with dimensionality smaller by one than the preceding one. Thus, after the $(l+1)$-fold execution of operation (8), we obtain set ϱ_{n+1} in the form of a single-point set which contains only one point $\overset{*}{L}$.

From the algebraic point of view, this result can be interpreted as follows. Assume that there are $(l+1)$ observations of form (1) with $f_n = 0$, i.e. assume that n takes values from $n = 1$ to $n = l$. Let us introduce the following designations:

$$Y_l = \begin{Vmatrix} y_1 \\ y_2 \\ \cdot \\ \cdot \\ \cdot \\ y_l \end{Vmatrix} , \qquad Z_l = \begin{Vmatrix} U_1^T \\ U_2^T \\ \cdot \\ \cdot \\ \cdot \\ U_l^T \end{Vmatrix} . \tag{10}$$

Then the considered system of observation equations can be written as

$$Z_l L = Y_l \ . \tag{11}$$

If the condition of linear independence of vectors U_n is met, i.e. that of linear independence of rows of matrix Z_1, then **det** $Z_1 \neq 0$ and we obtain from (11)

$$L = Z_1^{-1} Y_1 = \overset{*}{L} . \tag{12}$$

which is the required result.

Let us consider now a more interesting case when $f \neq 0$, i.e. $\Delta \neq 0$. Let us assume that there are two linearly-dependent observations among the whole set of observations (1), i.e.

$$U_s = cU_k , \tag{13}$$

where c is a known constant, such that

$$| f_s | = | f_k | = \Delta , \tag{14}$$

$$\operatorname{sign} f_s = -\operatorname{sign} f_k . \tag{15}$$

The fulfillment of conditions (14), (15) is in essence equivalent to the existence of one summary observation free from noise. Indeed, when condition (13) is met, we have

$$U_k^T L = y_k - f_k , \tag{16}$$

$$cU_k^T L = y_s + f_k . \tag{17}$$

From here we obtain that with fulfillment of (14), (15)

$$(1 + c) U_k^T L = y_k + y_s , \tag{18}$$

as we wished to prove.

It is obvious that the problem is how to single out the pair of observations (16), (17) sought for from the whole sequence of observation of type (1). Let us show that the application of procedure (8) allows to solve this problem.

Indeed, let conditions (13)-(15) be fulfilled and let $f_k = \Delta$ for definiteness. Then we obtain from (16) that one of two hypersurfaces determining the bounds of set $\tilde{\Omega}_k$, namely hypersurface

$$U_k^T L = y_k - \Delta$$

goes through point $L = \overset{*}{L}$. At the same time for observation of (17), the hypersurface bounding set $\tilde{\Omega}_s$ and determined by equation

$$cU_k^T L = y_s + \Delta$$

also goes through point $L = \overset{*}{L}$. The intersection of these two sets $\tilde{\Omega}_k$ and $\tilde{\Omega}_s$ (carried out in accordance with (8)) generates a set in the form of a hypersurface going through point $L = \overset{*}{L}$ (ref. to **Fig.** 2).

The existence of l pairs of such linearly independent observations as (16), (17) due to the successive execution of sets intersection operation (8) results in formation of a single-point set containing only one point $\overset{*}{L}$. All remaining observations from the sequence of observation (1) virtually do not take part in the determination of the final estimate and procedure (8) acts as a peculiar filter which screens all unnecessary observations.

In the general case, when the above pairs of observations may not exist in the required number, procedure (8) will determine some unimprovable estimate. In this case, a part of observations of form (1) appears to be non informative in the sense that it does not improve the available estimate of Ω_n.

Inasmuch as the check for fulfillment of conditions of form (13)-(15) in its labour content is in no way less than the realization of procedure (8) itself then it is obvious that it is more preferable to use the computing power of the computer directly for the realization of relationship (8).

As noted above, the fact that the sequence of estimates Ω_n generated by recurrent procedure (8) is a sequence of convex polyhedra allows to introduce a rather simple scalar estimate of the "quality" of the solution of the parametric identification problem. In doing so, we

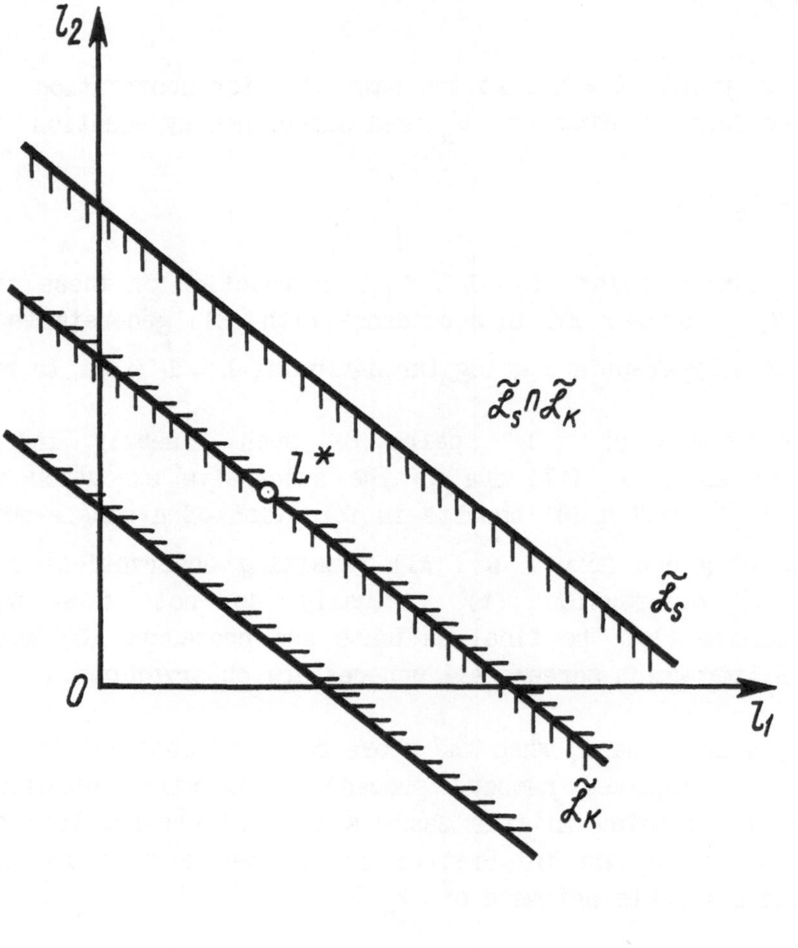

Fig. 2. Obtaining exact estimate in solving identification problem

shall proceed from the notion that "the less" is in some sense set Ω_n, the better (more exact) is solved the identification problem. To do this, it will suffice to introduce some scalar positive function $\delta(\Omega_n)$ defined over set Ω_n in the following way: if set Ω_n degenerates to a single-point set with $L = \overset{*}{L}$, then $\delta_n = \delta(\Omega_n) = 0$; if the "size" of set Ω_n grows then the value of this function should also grow.

It is obvious that if we take as function $\delta(\Omega_n)$ the so-called diameter of set Ω_n, i.e. the maximum distance between two vertices of polyhedron

$$\delta_n = \delta(\Omega_n) = \max_{i \in \overline{1,N}} \{ d_i(\Omega_n) \}, \tag{19}$$

where $d_i = d(\Omega_n)$ is the distance between any two vertices of set Ω_n, N is the number of vertices of the polyhedron then such function $\delta(\Omega_n)$ will satisfy the conditions stipulated above.

Let us note that just this abstract approach to the evaluation of the quality of the solution of the identification problem considering it as a completely independent problem irrespective of the main problem, is used as the basis for all known identification algorithms including those used in adaptive control systems. It is apparently justified only in the case when the identification and control problems are really being solved independently of one another.

Along with the "absolute" estimate of set Ω_n in form of its diameter, it is worthwhile to introduce also its "dimensionless" estimate in the form

$$\chi_n = \frac{\delta(\Omega_n)}{\delta(\Omega_n)},$$

which characterizes the relative measure of uncertainty of these sets. It is obvious that

$$0 \leqslant \chi_n \leqslant 1 \qquad \forall\, n > 0 .$$

Having available the introduced estimate (19) as the solution of the identification problem, we can terminate the attempts of its further refinement when it has attained some preset value $\overset{*}{\delta}$.

To determine set ϱ_{n+1}, we shall according to (8) execute the operation of intersection of sets ϱ_n with set $\tilde{\varrho}_{n+1}$ which is a certain "hyperstrip" which has according to (7) the following boundary equations:

$$U^T_{n+1}L = y_{n+1} + \Delta \ , \tag{20}$$

$$U^T_{n+1}L = y_{n+1} - \Delta \ . \tag{21}$$

Thus, when implementing in a computer the procedure of obtaining guaranteed estimates of the parameter vector, the main point is the construction of a formalized procedure of intersection of ϱ_n with halfspaces separated by linear inequalities (20) and (21). For the required description of the process in sufficient detail refer to [9]-[11] and here let us only note the following circumstance.

Obviously, the number of vertices of polyhedra ϱ_n is generally not equal to the number of vertices of polyhedron ϱ_n and, in particular, can exceed it. Because of this, when realizing the above algorithm for constructing sequence ϱ_n by means of a computer, this must be kept in mind and a required reserve of memory must be saved to store data array with variable dimensionality. It was found in an experimental test of the proposed algorithm that both an increase in the number of the vertices of polyhedra ϱ_n (due to the arising "new" vertices) and its decrease (due to the rejection of "old" vertices) takes place in the process of construction of sequence ϱ_n . This results in the situation when a linear increase in the number of vertices of the polyhedra is not observed with increasing n.

As an illustration of the use of the above method of constructing guaranteed estimates, let us consider the following example. Let $l=3$ in (1), i.e. (1) has the form

$$y_n = l_1u_{1n} + l_2u_{2n} + l_3u_{3n} + f_n \ ,$$

where true parameter values are:

$$\overset{*}{l}_1 = 1 \; ; \quad \overset{*}{l}_2 = 1 \; ; \quad \overset{*}{l}_3 = 2 \; .$$

The values of disturbance f_n have been selected arbitrarily from interval $[-1; 1]$. As a priori information about the parameter values, the estimates of the following were taken:

$$1 \leqslant l_1 \leqslant 5 \; ; \quad 1 \leqslant l_2 \leqslant 5 \; ; \quad 1 \leqslant l_3 \leqslant 3 \; .$$

Table 1 presents the sequences of values of given inputs u_{in} ($i = 1, 2, 3$) and of the measured output y_n (calculated in accordance with the true value of uncontrollable disturbance f_n also presented in **Table 1** and with the values of true parameters). Using these data, a procedure of identification of this plant was realized in a computer, in accordance with the technique described above for the case when $| f_n | \leqslant 1 \quad \forall n > 0$.

Presented in **Table 2** are data on actual change in the number of N_n after each step of identification based on a new measurement, as well as the changes in the "size" of set Ω_n , characterized by its diameter $\delta (\Omega_n)$. For a more illustrative presentation, same data are shown in graphical form in **Fig. 3** .

The analysis of the modeling results shows that the identification procedure ends in this case in a final number of steps (i.e., true parameter values are determined in this case) when disturbance f_n takes its boundary values for the both signs.

However, if real values $| f_n | < 1$, then the accuracy of its a priori estimate is deliberately lowered, the decrease of δ_n values decelerates and a determination of true values of the plant parameters appears to be impossible. The final result of application of the proposed identification procedure in these circumstances is the obtaining of some unimprovable estimate.

At a sertain stage of the problem solution, the lowering of the precision of the obtained estimate can prove to be advisable by approximating polyhedron Ω_n "from above" with some other polyhedron of a simpler form (defined by a small number of inequalities - faces)

Table 1

n	1	2	3	4	5	6	7	8	9	10	11	12	13	14
u_{1n}	0.033	0.479	-0.316	0.619	-0.843	-0.697	0.682	0.269	0.379	0.896	-0.24	-0.299	-0.775	-0.154
u_{2n}	0.164	0.397	0.421	-0.903	-0.214	0.512	-0.59	-0.653	-0.105	0.748	0.798	0.506	0.126	-0.771
u_{3n}	0.819	-0.013	0.105	-0.514	-0.932	0.587	-0.949	0.735	-0.524	0.39	-0.012	0.529	0.634	0.143
y_n	1.931	0.788	0.838	-1.879	1.469	1.932	-2.552	0.763	-1.395	2.106	0.473	1.91	-0.211	0.073
f_n	0.096	-0.063	0.524	-0.568	0.661	0.936	-0.746	-0.324	-0.621	-0.048	-0.06	0.645	-0.831	0.714

Table 2

n	1	2	3	4	5	6	7	8	9	10	11	12	13	14
N_n	8	10	8	10	12	10	12	12	14	14	12	14	14	12
δ_n	6	6	3.65	3.58	3.42	2.48	2.18	2.18	2.05	2.05	1.95	1.61	1.61	1.55

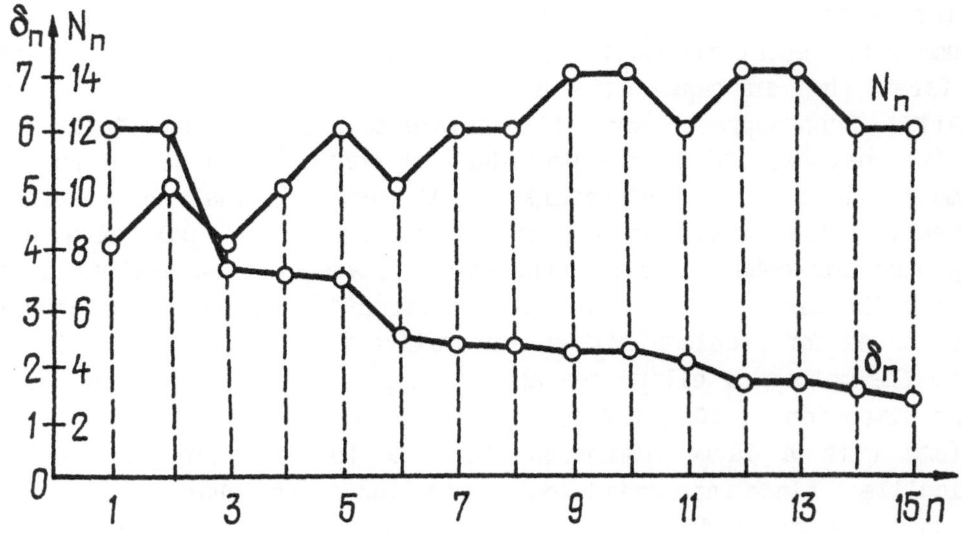

Fig. 3. Changing diameter of set $\mathcal{P}_n$ and number of its vertices

which can be taken to be some a priori estimate which should be refined later on.

Although the proposed procedure requires an increase in the volume of computations and in the required main memory capacity as compared with the traditional methods of parametric identification, however it repays itself by a larger body of information obtained as a result of the solution of the problem being set.

It is appropriate to note here the following. The set-theoretic approach to the solution of problem of obtaining guaranteed estimates in the form of estimates with lowered precision - ellipsoids circumscribed about respective polyhedra was apparently proposed for the first time in paper [12] and subsequently developed in papers [13]-[17]. Such approach makes it possible to simplify essentially the computational aspects of the procedure of construction of guaranteed estimates, however, unfortunately, at the cost of their coarsening. Apparently, the areas of efficient application of polyhedra and ellipsoids for constructing guaranteed estimates can be evaluated as follows. The exact method, i.e. the construction of polyhedra should be preferred for problems with a comparatively small dimension ($l \leq 10$) . The method of ellipsoids should be preferred for problems with medium dimension ($10 \leq l \leq 10^2$) . The use of the both methods for problems with a large dimension ($l > 10^2$) involves severe difficulties since approximation error increases sharply for the ellipsoid method and the required memory space and volume of computations grow to an inadmissible extent.

In the foregoing, we have considered the simplest case of the parametric identification problem when the uncontrollable disturbance was an additive part of the measured value. In this case, the identification problem was practically redused to the problem of finding the set of solutions of a respective system of linear algebraic equations given uncertainty only in its right-hand side. The situation changes substantially when an uncontrollable disturbance is applied to the input of the system being studied.

Let us consider now a more general class of systems than that described by equation (1). Let a class of plants be given described by equation

$$y_n = L^T(U_n + \tilde{F}_n) + f_n , \qquad n = 1, 2, 3, \ldots, N . \qquad (22)$$

where all designations (except for $\tilde{F}_n$) have the same meaning as in
(1) and $\tilde{F}_n$ is 1-dimensional vector of uncontrollable disturbances
with a priori estimate

$$\tilde{F}_n \in \tilde{\mathfrak{F}} \qquad \forall n > 0 . \qquad (23)$$

where $\tilde{\mathfrak{F}}$ is a given bounded convex set (polyhedron).

Let a priori estimate (2) be given for unknown parameter vector L.
It is required to construct a reccurent procedure for refinement of
$\mathfrak{L}_0$ using measured data U_n , y_n , equation (22) and estimates (3)
and (23).

General plan for the solution of the parametric identification
problem remains the same as presented above. Let us assume that there
was estimate $L \in \mathfrak{L}_n$ at the n-th step of observation of system (22).
Then we obtain from equation (22) for the (n+1)-th step:

$$L \in \tilde{\mathfrak{L}}_{n+1} = \{ L| (U_{n+1} + \tilde{F}_{n+1})^T L + f_{n+1} - y_{n+1} = 0 \}$$

and then we obtain a new a posteriori estimate when executing the
operation of intersection of sets $\tilde{\mathfrak{L}}_{n+1}$ and $\mathfrak{L}_n$ in accordance with
(8). The availability of unknown vector function $\tilde{F}_n$ bounded only by
its a priori estimate (23) presents substantial problems both in
constructing set $\tilde{\mathfrak{L}}_{n+1}$ itself and in realizing the operation of its
intersection with set $\mathfrak{L}_n$. Because of this, let us dewell on the
matter in greater detail (ref. to [18], [11]).

Denote

$$A_n^T = U_n^T + \tilde{F}_n^T , \qquad b_n = y_n - f_n , \qquad (24)$$

$$A = \begin{Vmatrix} A_1^T \\ A_2^T \\ \cdot \\ \cdot \\ \cdot \\ A_N \end{Vmatrix}, \qquad B = \begin{Vmatrix} b_1 \\ b_2 \\ \cdot \\ \cdot \\ \cdot \\ b_N \end{Vmatrix}. \tag{25}$$

Using (24) and (25) as well as given estimates (3) and (22), we can evidently determine estimates A and B and then we obtain

$$A \in \mathfrak{A}, \quad B \in \mathfrak{B}. \tag{26}$$

Introducing re-designation $X = L$ and, respectively, instead of (2)

$$X \in \mathfrak{X}_0 ,$$

we can write system of linear algebraic equation (20) in "canonical" form

$$AX = B . \tag{27}$$

The problem of searching for efficient methods of solving a system of linear algebraic equation in different formulations is one of ancient problems in mathematics. The presence of unavoidable errors (inaccuracies) in setting coefficients both in the right-hand and in the left-hand side of the system arising either from the inaccuracy of the initial data themselves in that meaningful problem whose mathematical model is the system of equations being considered, or from the final accuracy of numarical representation in a computer or from the both results in indefinitness of the desired solution. When varying the coefficients of a system of equations within the accuracy of their setting, we can obtain different solutions equally pretending to be the true solution of the initial meaningful problem. However, the existing methods for solving linear equation systems like, e.g., the least-squares method, the Tikhonov's regularization method etc. are oriented to obtain only the unique solution and they are the concrete manifestation of that "... subjective aversion to problems having no single-valued answer" about which Kalman spoke in [19].

In the last few years, a number of problems have been studied like, for example, the problems of control, identification and filtering under conditions of uncertainty, which are not of stochastic nature and require for their solution the determination og the whole set of solutions of a system of linear equations with indeterminacy in the values of coefficients in both sides of this system. In a similar statement, such problems are being considered by the supporters of a new trend in science which was named the "interval mathematics" (e.g., ref. to [20]-[22]).

According to definitions of function transforms over sets, let us define the function (mapping) of elements $\mathbf{A} \in \mathfrak{A}$ and $\mathbf{B} \in \mathfrak{B}$ ([23], [10]):

$$\mathfrak{F}(\mathbf{A},\mathbf{B}) = \{ \mathbf{X} \mid \mathbf{AX} = \mathbf{B} \} ,$$

where $\mathfrak{F}(\cdot)$ is in the general case some set in $\mathbf{R}^1$ (in particular a single-point one for the case $1 = \mathbf{N}$ and nonsingular matrix $\mathbf{A}$).

This set is empty in the event that there exist no $\mathbf{X}$ satisfying (1) for the given $\mathbf{A}$ and $\mathbf{B}$.

Let us look for set

$$\mathfrak{X} = \mathfrak{F}(\mathfrak{A},\mathfrak{B}) = \bigcup_{\mathbf{A}\in\mathfrak{A}} \bigcup_{\mathbf{B}\in\mathfrak{B}} \mathfrak{F}(\mathbf{A},\mathbf{B}) , \qquad (28)$$

i.e. for the set of all feasible solutions of the system of equations (27) with all possible according to (26) values of coefficients of matrix $\mathbf{A}$ and the right-hand sides of the system. In other words, let us give the following

<u>Definition 1.</u> The solution of system (27) and (26) is taken to mean the set $\mathfrak{X}$ of all possible $\mathbf{X} \in \mathbf{R}^1$ for each of which such $\mathbf{A}$ and $\mathbf{B}$ satisfying (26) will be found that (27) will be satisfied.

It is natural that generally, in the absence of additional conditions of selection all elements of this set (if it is not empty) are equal in rights in the sense that none of them can pretend to be the unique "correct" solution.

Let us note that in paper [24] the solution was meant in narrower sense, namely as a set of nonsingular matrices $A^{-1}B$.

Let us consider a special case when set $\mathfrak{A}$ is a single-point one, i.e. $A = \mathfrak{A}$, where A is a given number matrix. Then for the case ofsquare ($l = N$) nonsingular matrix A we obtain

$$\mathfrak{F}(A,B) = A^{-1}B$$

and

$$\mathfrak{X} = \bigcup_{B\in\mathfrak{B}} (A^{-1}B) .$$

In this case, set $\mathfrak{X}$ is a linear transformation of set $\mathfrak{B}$. To take an example, when $\mathfrak{B}$ is a convex polyhedral set in space R^l with given vertices B^j, then $\mathfrak{X}$ will be also a convex polyhedron in R^l, with coordinates of vertices X^j equal to

$$X^j = A^{-1}B^j .$$

Let us consider now a more general case when

$$\mathfrak{B} = \bigcup_r \mathfrak{B}_r , \quad r \in \overline{1,N_B} , \tag{29}$$

where $\mathfrak{B}_r$ are some subsets of set $\mathfrak{B}$, N_B is the number of these subsets and

$$\mathfrak{A} = \bigcup_k \mathfrak{A}_k , \quad k \in \overline{1,N_A} , \tag{30}$$

where $\mathfrak{A}_k$ are some subsets of set $\mathfrak{A}$, N_A is the number of these subsets. Then substituting (29) and (30) into (28), we obtain

$$\mathfrak{X} = \bigcup_k \bigcup_r \mathfrak{X}_{kr} , \quad k \in \overline{1,N_A} , \quad r \in \overline{1,N_B} , \tag{31}$$

where

$$\mathfrak{X}_{kr} = \bigcup_{A\in\mathfrak{A}_k} \bigcup_{B\in\mathfrak{B}_r} \mathfrak{F}(A,B) . \tag{32}$$

It follows from inequalities (31) and (32) that set $\mathfrak{X}$ sought for is the solution of linear equations system (27) under the condition (26) in the sense of the above definition and it possesses a property which we shall call the property of compositiveness, viz. this set can be obtained from some subsets of form (32) by their union. This means, that in determining set $\mathfrak{X}$, we can represent sets $\mathfrak{A}$ and $\mathfrak{B}$ in form of (28) and (29), i.e. in form of a union of some number of subsets of standard form so that a set of form (32) can be determined for each rair of the sets, and whereupon we can use relationship (31). It follows from the foregoing that it is necessary to have a constructive procedure for determining the whole set of solutions of system (27) for some standard form of sets $\mathfrak{A}$ and $\mathfrak{B}$.

Consider the problem of determining the solution of system (27) (in the sense of the above definition) for the case when the estimates are known of matrix A and of the right-hand sides of the equations in form

$$\underline{a}_{nj} \leqslant a_{nj} \leqslant \bar{a}_{nj}, \quad n \in \overline{1,N}, \quad j \in \overline{1,I}, \tag{33}$$

$$\underline{b}_n \leqslant b_n \leqslant \bar{b}_n, \quad n \in \overline{1,N}, \tag{34}$$

where $\underline{a}_{nj}$, $\bar{a}_{nj}$ and $\underline{b}_n$, $\bar{b}_n$ are given numbers. The availability of estimates (33) and (34) means in terms of the "interval mathematics" [22], [24] that matrix A and vector B in equation (27) are "interval" ones, i.e. given are not A and B themselves, but only the intervals to which their elements belong.

In the foregoing, a constructive procedure was described for determining a set of solutions of system (27) under "interval" constrains (34) imposed on the elements of vector B and with known values of matrix A, i.e. corresponding to the case, when $\underline{a}_{nj} = \bar{a}_{nj}$ in (33). In this case the solution is a convex set - a polyhedron. However, if there is uncertainty in the velues of elements of matrix A, the situation is more complex. It is easy to verify that generally set $\mathfrak{X}$ will be non-convex. Indeed, let us consider, for example, one equation of system (27) with two-dimensional vector X, i.e. with $I = 2$, $N = 1$ (two-dimensional case is convenient for geometric interpretation of the results on a plane). Qualitative view of set $\mathfrak{X}$ is shown in **Figs. 4 - 6** (crosshatched area) for $\underline{b}_1 > 0$ and different values of elements a_{11} and a_{12} . A case with $\underline{a}_{11} > 0$,$\underline{a}_{12} > 0$ is shown in **Fig. 4** ; a case with $\underline{a}_{11} < 0$, $\bar{a}_{11} > 0$, $\underline{a}_{12} > 0$ is shown in **Fig. 5** ; a case with $\underline{a}_{11} < 0$, $\bar{a}_{11} > 0$, $\bar{a}_{12} > 0$, is shown in **Fig. 6** .

Let us consider the s-th orthant $(s = 1, \ldots, 2^l)$ of the system solution space R^l . Let us define the set of indices $e_j^s = 0$ for this orthant with $j = 1, \ldots, l$ in the following way: $e_j^s = 0$ when the value of component x_j of vector X is positive in this orthant; $e_j^s = 1$ in the opposite case. Then square matrix $G_s = \mathbf{diag}\,\{e_1^s, \ldots, e_l^s\}$ will characterize the s-th orthant being considered and this orthant is isolated by inequality

$$(I - 2G_s)X \geq 0 . \tag{35}$$

Let us introduse matrices $\underline{C}_s(\cdot)$ and $\bar{C}_s(\cdot)$ in each s-th orthant with elements defined as follows:

$$\underline{c}_{nj}(s) = a_{nj}^{e_j^s} , \qquad \bar{c}_{nj}(s) = a_{nj}^{1-e_j^s} , \quad n \in \overline{1,N} , \quad j \in \overline{1,l} . \tag{36}$$

Let us introduse also vectors

$$\underline{B}^T = (\underline{b}_1 , \ldots, \underline{b}_N) , \qquad \bar{B}^T = (\bar{b}_1 , \ldots, \bar{b}_N) .$$

The system of nonlinear inequalities

$$\left.\begin{aligned} \underline{C}_s X &\leq \bar{B} , \\ \bar{C}_s X &\geq \underline{B} \end{aligned}\right\} \tag{37}$$

jointly with inequality (35) extracts set $\mathfrak{X}^s$ which is a convex polyhedron.

True is the following

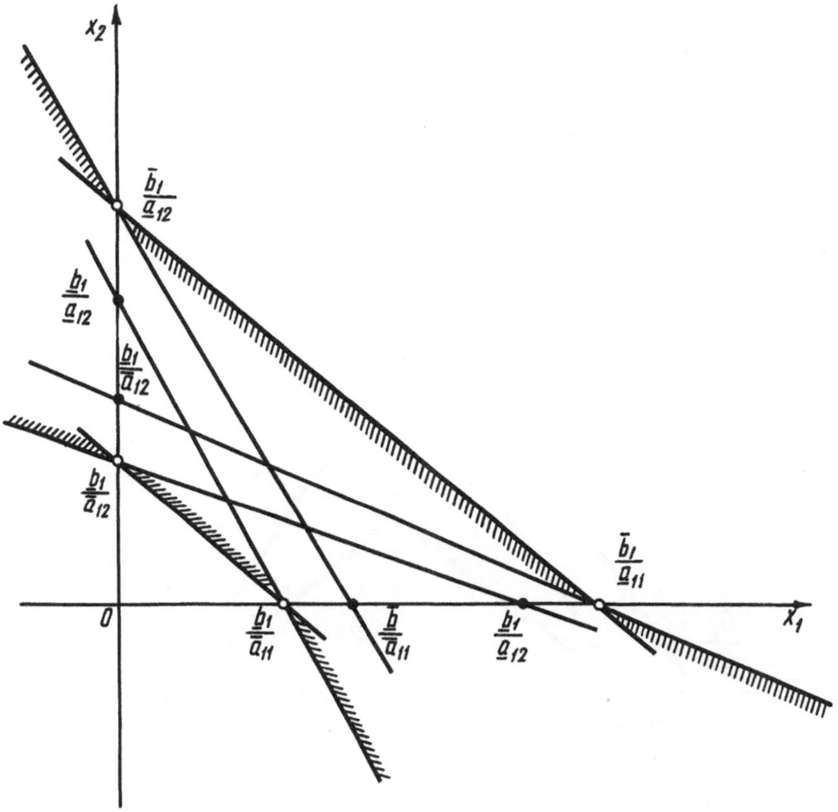

Fig. 4. Set of solutions of linear equation with $\underline{a}_{11}>0$, $\bar{a}_{12}>0$

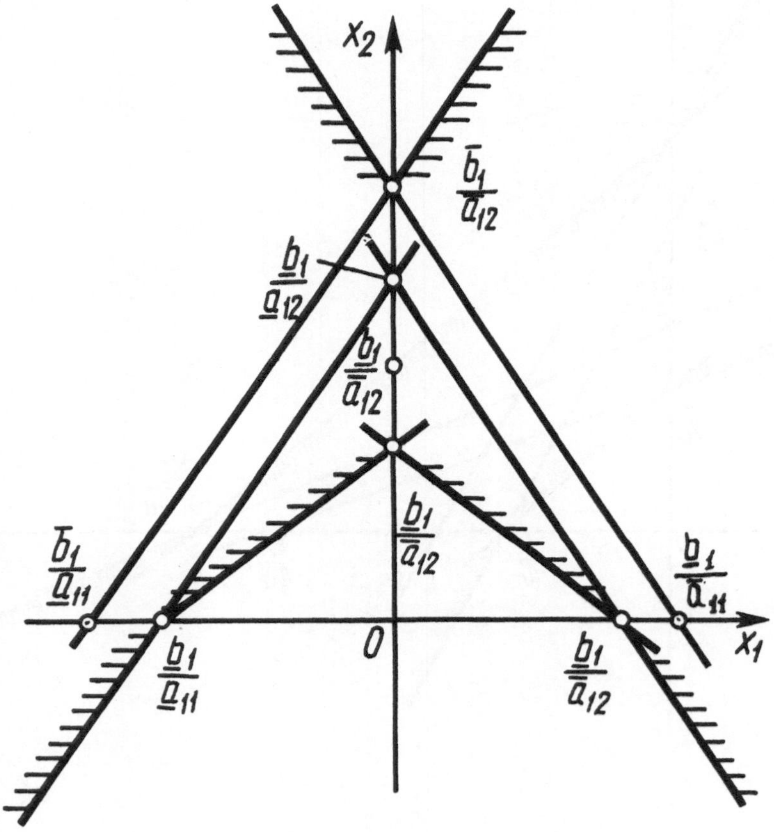

Fig. 5. Set of solutions of linear equation with $\underline{a}_{11}<0$, $\overline{a}_{11}>0$, $\underline{a}_{12}>0$

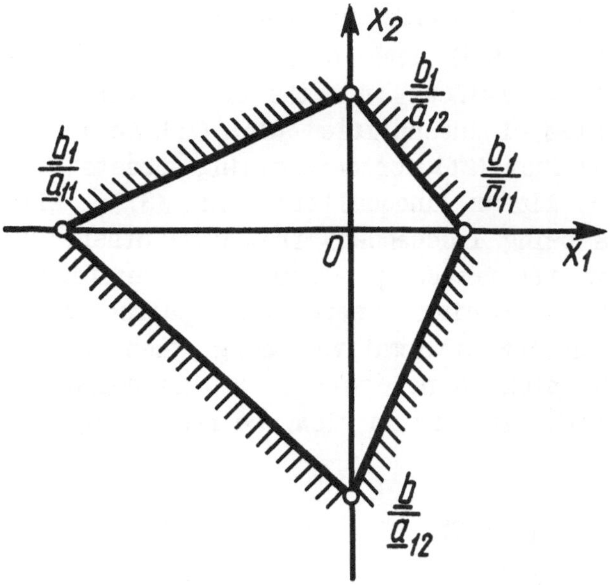

Fig. 6. Set of solutions of linear equation with $\underline{a}_{11}<0$, $\overline{a}_{11}>0$, $\underline{a}_{12}>0$

Theorem 1. The set of solutions of system (27) under conditions (26) which have form (33), (34) is the union of convex polyhedra $\mathfrak{X}^s$ over all orthants

$$\mathfrak{X} = \bigcup_{s=1}^{2^l} \mathfrak{X}^s . \qquad (38)$$

The proof of the theorem is presented in [18] and we omit it here.

Naturally, some of subsets $\mathfrak{X}^s$ can turn out to be empty, i.e. system of inequalities (35) and (36) for some orthants proves to be inconsistent. It follows from the comparison of system of equations (27) and system of inequalities (37) (taking into account (36) that each equation from (27) for determining subsets $\mathfrak{X}^s$ generates in each orthant two linear inequalities in (37), all these pairs of inequalities being independent from each other. This property allows to construct recurrence procedure for refining subsets $\mathfrak{X}^s$ by successively adjoining respective pairs of inequalities and eliminating the non-informative among them (i.e. those pairs of inequalities which do not refine $\mathfrak{X}$ as compared with the previous estimate). With this aim in view, we rewrite system of equations (27) in the form:

$$A_n^T X = b_n , \quad n \in \overline{1,N} , \qquad (39)$$

where A_n^T is the n-th row of matrix A . Set $\tilde{\mathfrak{X}}_n$ corresponds to each scalar equation from (39), such that for each $X_n \in \tilde{\mathfrak{X}}_n$ there will be found $A_n \in \mathfrak{A}$ and $b_n \in \mathfrak{b}$ satisfying equality (39) for selected n . According to Teorem 1 , each set $\tilde{\mathfrak{X}}_n$ can be represented in the form:

$$\tilde{\mathfrak{X}}_n = \bigcup_{s=1}^{M} \tilde{\mathfrak{X}}_n^s , \quad M = 2^l , \qquad (40)$$

where $\mathfrak{X}_n$ is a convex set completely belonging to the s-th orthant of space R^l and isolated in this orthant by two scalar inequalities

$$\left.\begin{aligned} &\bar{C}^T_{ns} X - \underline{b}_n \geqslant 0 \,, \\ &\underline{C}^T_{ns} X - \bar{b}_n \leqslant 0 \,, \end{aligned}\right. \tag{41}$$

where $\underline{C}^T_{ns}$ and $\bar{C}^T_{ns}$ - are n-th rows of matrices $\underline{C}_s$ and $\bar{C}_s$.

On the other hand, by virtue of the independence of restrictions imposed upon different elements of matrix **A** and vector **B** , we obtain that

$$\mathfrak{X} = \tilde{\mathfrak{X}}_1 \cap \tilde{\mathfrak{X}}_2 \cap \cdots \cap \tilde{\mathfrak{X}}_N \,. \tag{42}$$

Difference equation of evolution of sets follows from here

$$\mathfrak{X}_{n+1} = \tilde{\mathfrak{X}}_{n+1} \cap \mathfrak{X}_n \,, \quad \mathfrak{X}_1 = \tilde{\mathfrak{X}}_1 \,, \quad n \in \overline{1,N} \,, \tag{43}$$

such that

$$\mathfrak{X} = \mathfrak{X}_N \,. \tag{44}$$

From (40) and (43), we obtain a system of independent difference equations

$$\left.\begin{aligned} &\mathfrak{X}^i_{n+1} = \tilde{\mathfrak{X}}^i_{n+1} \cap \mathfrak{X}^i_n \,, \quad i \in \overline{1,M} \,, \quad n \in \overline{1,N} \,, \\ &\mathfrak{X}^i = \tilde{\mathfrak{X}}^i \,, \end{aligned}\right\} \tag{45}$$

which describe the evolution of imbedded convex polyhedra in each orthant of space R^1 separately (with the exception of those orthants where these sets turn to be empty). When this takes place, the procedure of realization in each equation from (45) of the operation of intersection of convex sets $\mathfrak{X}^i_n$ and $\tilde{\mathfrak{X}}^i_{n+1}$ remains similar to the procedure described above, since in case of boundness of sets $\mathfrak{X}^i_n$ it is necessary only to realize the intersection of the polyhedron with half-spaces (41) and the "rejection" of the cut off parts.

Let us note that with a priori estimate of set $\mathfrak{X}$ in form

$$\mathfrak{X} \subset \mathfrak{X}_0 , \tag{46}$$

where $\mathfrak{X}_0$ is a given set represented as a union of convex polyhedra $\mathfrak{X}_0^i$ belonging to respective orthants

$$\mathfrak{X}_0 = \bigcup_i \mathfrak{X}_0^i , \quad i \in \overline{1,M} , \tag{47}$$

such estimate can be constructively used in the described recurrent procedure. To do this requires no more than to extend equation (45) also to the value $n = 0$ by replacing initial condition $\mathfrak{X}_1^i = \tilde{\mathfrak{X}}_1^i$ by condition (47).

Let us illustrate the above procedure of finding the set of solutions of system of equations (27) under conditions (26) using the following example.

Consider the problem of the set identification of a static plant described by the equation

$$y_n = (U_n + V_n)^T X + w_n , \quad n \in \overline{1,N} , \tag{48}$$

where n is the measurement number, y_n is the measured scalar output, $U_n^T = (u_{1n}, \ldots, u_{ln})$ is the vector of output signals, $X^T = (x_1, \ldots, x_l)$ is the vector of indeterminate parameters of plants, $V_n^T = (v_{1n}, \ldots, v_{ln})$ is noise satisfying the condition

$$| v_{jn} | \leqslant \Delta_j , \quad j \in \overline{1,l} ; \qquad | w_n | \leqslant \tilde{\Delta} , \quad n \in \overline{1,N} . \tag{49}$$

It is required to estimate the true value of parameter vector X on the basis of all N measurements. The problem stated in this way obviously can be reformulated and reduced to the problem considered above.

Indeed, a set of N linear relationships (48) obtained as a result of measurements comprises a system of equations of the form (27). When this takes place, estimates (33) and (34) of the coefficients of matrix A and vector B have the form in accordance with (48) and (49):

$$u_{jn} - \Delta_j \leqslant a_{jn} \leqslant u_{jn} + \Delta_j \,, \qquad (50)$$

$$y_n - \tilde{\Delta} \leqslant b_n \leqslant y_n + \tilde{\Delta} \,. \qquad (51)$$

Then the set identification problem can be formulated as the problem of finding a set solution of linear equation system (27) under conditions (50) and (51) in accordance with the foregoing definition. It is obvious that set $\mathfrak{X}$ obtained in this way will comprise the unknown true value of the plant parameter vector (48) and some its vicinity, i.e. it will just represent the estimate of this value.

A special case of system (27) with $l = 2$, $N = 4$, under conditions (50) and (51) with $\tilde{\Delta} = 4$, $\Delta_1 = \Delta_2 = 0{,}4$ was simulated in a computer. In digital simulation of static plant (48), the value of parameters vector X unknown for the investigator was assumed to be equal to (4;1) and noise v_{jn} and w_n was generated by random-number generators with uniform distribution within the above range. The following values of input U_n where used: (1;2), (2;-1), (1;-2), (2;1). In this case, measurements y_n represented, respectively: 5.0; 6.2; 2.4; 7.5 .

With initial data specified above, the set solution of the respective system of equations was found, i.e. systems of inequalities describing sets $\mathfrak{X}^i$, $i \in \overline{1,4}$ in each i-th quadrant have been obtained. Using a special program, the solutions of these systems of inequalities (for non-empty sets $\mathfrak{X}^i$) have been found in the form of coordinates of vertices of convex polyhedra.

Let us describe the complete set of solutions $\mathfrak{X}$. **Fig.** 7 shows intersection of two regions $\tilde{\mathfrak{X}}_1$ and $\tilde{\mathfrak{X}}_2$ generated by the pairs of inequalities corresponding to the 1st and 2nd equations in system (27) under conditions (50), (51). **Figs.** 7,8 shows how this intersection jointly with two remaining regions $\tilde{\mathfrak{X}}_3$ and $\tilde{\mathfrak{X}}_4$ (corresponding to the 3d and 4th equations) defines set $\mathfrak{X}$ sought for. As a result, we obtain

$$\mathfrak{X} = \mathfrak{X}^1 \cup \mathfrak{X}^2 \,,$$

where subset $\mathfrak{X}^1$ in the first ($i = 1$) quadrant represents a heptagon

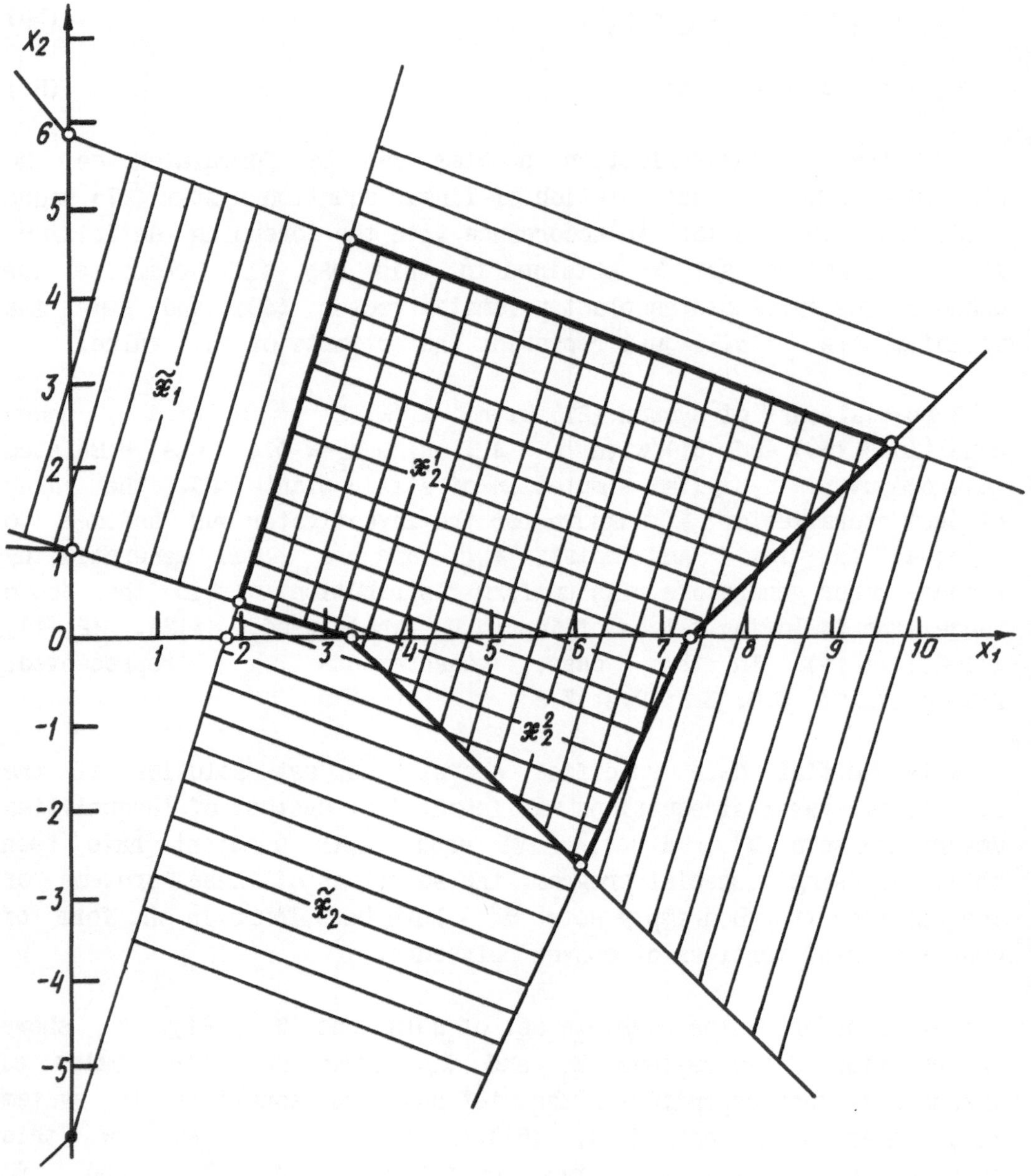

Fig. 7. Set of solutions of a system of two linear equations

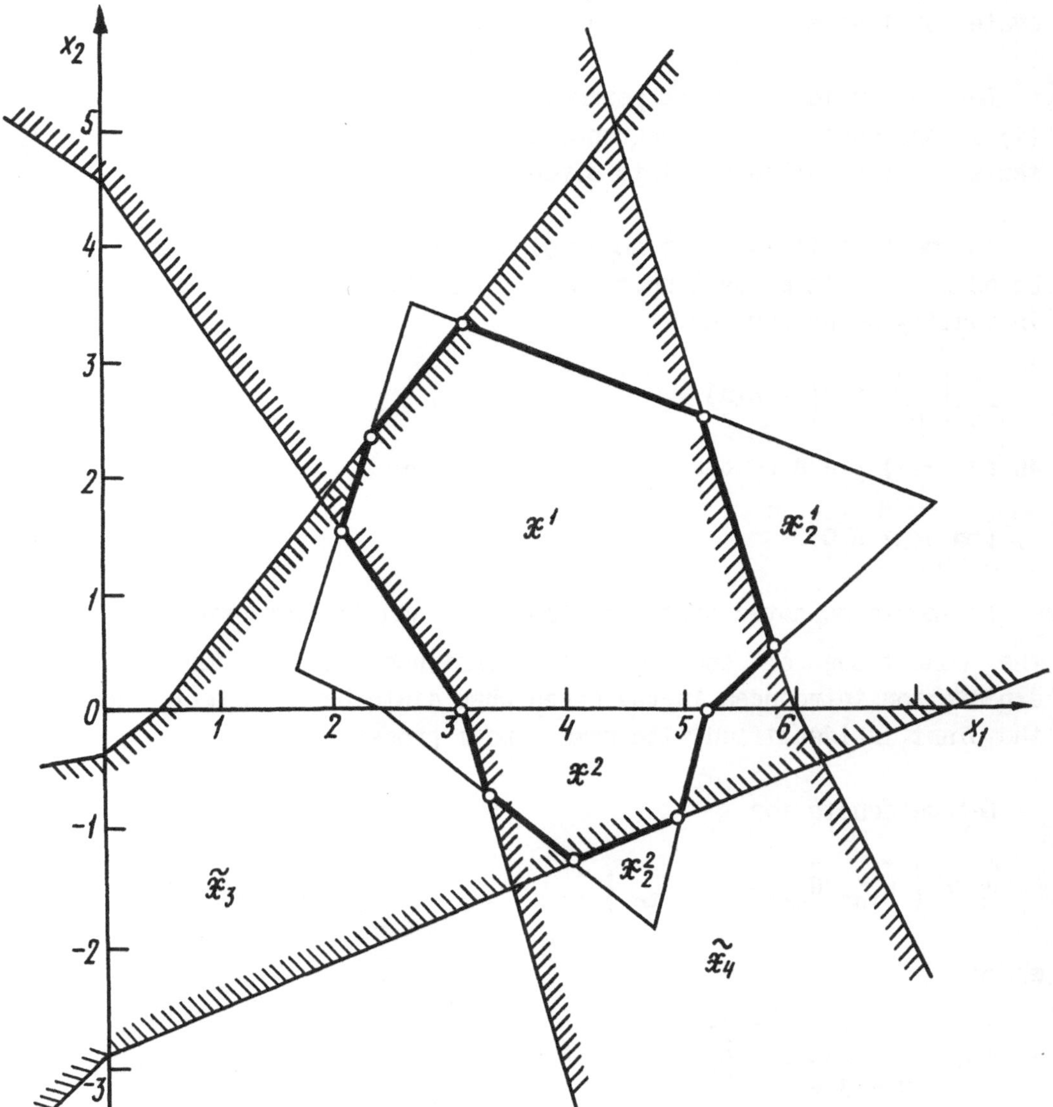

Fig. 8. Refining the set of solutions of a system of two linear equations using a priori estimate

and is described by 10 inequalities, 7 of them being informative and 3 not; subset $\mathfrak{X}^2$ in the second (i = 2) quadrant constitutes a pentagon also described by 10 inequalities, 5 of them being informative. The remaining subsets $\mathfrak{X}^4$ and $\mathfrak{X}^3$ are empty since the system of inequalities defining them are incompatible.

Let us consider now the parameter identification problem for system (1) in the case that the sequence of uncontrollable disturbances f_n meets a system of constraints stronger than (3).

Assume that it is known about disturbance f_n not only that it is bounded, i.e. it meets condition (3), but what is more, the following inequality holds for any

$$\frac{1}{n}\left| \sum_{k=1}^{n} f_n \right| \leqslant \sigma(n) \,, \tag{52}$$

where $\sigma(\cdot)$ is a bounded function, such that

$$\lim_{n \to \infty} \sigma(n) = 0. \tag{53}$$

It should be noted that conditions (3) and (53) are similar to the conditions of boundness of variance and mathematical expectation introduced in the cases when disturbance acting upon the plant is identified with some random process.

Let us denote for

$$\tilde{U}_n = \left| \bar{U}_{1n}, \bar{U}_{2n}, \ldots, \bar{U}_{ln} \right| , \tag{54}$$

where

$$\bar{U}_{jn} = \frac{1}{n - j + 1} \sum_{k=1}^{n-j+1} U_k \,, \quad j \in \overline{1,l} \,. \tag{55}$$

Then true is the following

<u>Statement 1.</u> When input actions for plant (1) are such that, beginning with some sufficiently large $n > 1$, matrix $\tilde{U}_n$ defined by expressions (54) and (55) is non-singular and the external

non-controllable disturbance f_n satisfies conditions (4), (52) and (53), then the identification procedure described above allows with increasing number of measurements n to obtain arbitrarily accurate estimate of parameter vector L in the sense that linear dimensions of set Ω_n (e.g., its projections onto coordinate axes of space {L}) will be arbitrarily small.

To prove the truth of this statement, let us derive from (1) with n > l the equations of the form

$$\frac{1}{n-j+1}\sum_{k=1}^{n-j+1} y_k = \frac{1}{n-j+1}\left[\left(\sum_{k=1}^{n-j+1} U_k^T\right)L + \sum_{k=1}^{n-j+1} f_k\right],\ j \in \overline{1,l}.$$

Let us write this system of equations in vector-matrix form

$$\bar{Y}_n = \tilde{U}_n^T L + \bar{F}_n\ , \quad n = l+1,\ l+2,\ \dots,$$

where

$$\bar{Y}_n^T = (\ \bar{y}_{1n},\ \bar{y}_{2n},\ \dots,\ \bar{y}_{ln})\ ,$$

$$\bar{y}_{jn} = \frac{1}{n-j+1}\sum_{k=1}^{n-j+1} y_k\ .$$

$$\bar{F}_n^T = (\ \bar{f}_{1n},\ \bar{f}_{2n},\ \dots,\ \bar{f}_{ln})\ ,$$

$$\bar{f}_{jn} = \frac{1}{n-j+1}\sum_{k=1}^{n-j+1} f_k\ .$$

Since with rather large n , the constraints on the absolute value of components $\bar{f}_{jn}$ of vector $\bar{F}_n$ tend to zero, then the truth of **Statement** 1 follows from the non-singularity of matrix $\tilde{U}_n$ and relationships (6), (7) and (8).

Let us note that in practical realization of the procedure constructing the sequence of estimates Ω_n from data of newly arriving measurements, the current guaranteed estimate Ω_n is determined as the intersection of the estimate Ω_{n-1} available earlier with two sets: $\check{\Omega}_n$ and $\hat{\Omega}_n$ - "hyperstrips" in space {L} defined by the respective inequalities:

$$| y_n - U_n^T L | \leqslant \Delta ,$$

$$| \bar{y}_{jn} - \bar{U}_{jn}^T L | \leqslant \sigma(n) .$$

Each of these inequalities can be presented in the form of two linear inequalities, i.e.

$$\left.\begin{aligned} & y_n - U_n^T L - \Delta \leqslant 0 , \\ & - y_n + U_n^T L - \Delta \leqslant 0 , \\ & \bar{y}_{jn} - \bar{U}_{jn}^T L - \sigma(n) \leqslant 0 , \\ & -\bar{y}_{jn} + \bar{U}_{jn}^T L - \sigma(n) \leqslant 0 . \end{aligned}\right\} \qquad (56)$$

The first two inequalities from system (56) define with each n a polyhedron $\hat{\varrho}_n$ whose intersection with polyhedron ϱ_{n-1} is carried out in accordance with the procedure described above. The intersection of this newly obtained polyhedron with polyhedron $\check{\varrho}_n$ defined by the third and fourth inequalities of system (56) is carried out also in accordance with the procedure described above.

As already mentioned above, so far we have considered the passive identification case since it was assumed that the choice of the control sequence U_n was specified in some way and did not depend on the researcher. Let us consider now the case of active identification which is taken to mean a selection of control sequence which provides an optimal in some sense quality of the solution of this problem.

Thus, as before let us consider plant being studied (1). In the foregoing, it was found that sets ϱ_n are convex polyhedra and therefore their diameter δ_n is

$$\delta_n = \delta(\varrho_n) = \max_{j,k} | L_n^j - L_n^k | \qquad (57)$$

where L_n^k is the k-th vertex of polyhedron ϱ_n . In so doing, let us assume that diameters of the empty and the single-point sets are equal to zero.

Having some estimate $\mathfrak{P}_{n-1}$ prior to the n-th step of control (and measurement), it is advisable to select control U_n reasoning from the minimization of diameter (57). In this case, it is taken into account that according to (7) and (8) diameter of set $\mathfrak{P}_n$ will be an explicit function of control U_n and of the measurement result y_n not known in advance. Let us describe the set of values $\mathfrak{Y}_n$ to which belongs the next measurement y_n due depending on the set control U_n, on the set of disturbances $\mathfrak{f}$ and the set of values of the vector of uncertain parameters. It has the form

$$\mathfrak{Y}_n = \mathfrak{P}_{n-1} \times U_n + \mathfrak{f} \ .$$

The operations of multiplication and summation are taken here to mean the sets of respective sums and products with arguments assuming arbitrary values from the above sets. Using game approach for the completion of the definition of the properties of the vector of uncertain parameters and disturbances, let us look for the solution of the active identification problem in the form

$$\min_{U_n \in \mathfrak{U}} \max_{y_n \in \mathfrak{Y}} \{ \delta [\mathfrak{P}_n(y_n, U_n)] \} \ . \tag{58}$$

An important property of optimal control U_n takes place for the case of convex set $\mathfrak{U}$, namely: it assumes its values at the boundary $\bar{\mathfrak{U}}$ of this set. Actually, diameter of set $\mathfrak{P}_n$ depends according to (7) and (8) both on the orientation in space of hyperstrip $\tilde{\mathfrak{P}}_n$ and on its "thickness" ρ (distance between hypersurfaces bounding it) which is determined from (7) by the expression

$$\rho = 2\Delta \ / \ | \ U_n \ |^2 \ .$$

From this it follows that for each inner point $\tilde{U} \in \mathfrak{U} \setminus \bar{\mathfrak{U}}$ such vector $\check{U}_n \in \bar{\mathfrak{U}}$ can be found of the form

$$\check{U}_n = \alpha \tilde{U} \ , \quad \text{where} \quad | \ \alpha \ | > 1 \ ,$$

that with $U_n = \check{U}_n$ the thickness of the obtained hyperstrip will be less than for $U_n = \tilde{U}_n$ with one and the same orientation in the parameter space. Thus, valid is the following

<u>Lemma 1.</u> If set Ω in (5) is convex, then the solution of problem (58) coincides with the solution of problem

$$\min_{U_n \in \bar{\Omega}} \max_{y_n \in \mathfrak{Y}_n} \{ \, \delta[\, \mathfrak{L}_n(y_n, U_n) \,] \, \} \, . \tag{59}$$

Besides, it is possible to simplify the maximization with respect to y_n on set $\mathfrak{Y}_n$ when solving problem (59) for the selected value of U_n . Assume that there is polyhedron $\mathfrak{L}_{n-1}$ with vertices L_{n-1}^T , $k = 1, \dots, N_{n-1}$ where N_{n-1} is the number of these vertices. Denote a set in the parameter space of the form

$$\tilde{\mathfrak{L}}_n^k = \{ \, L: \, | \, U_n^T(L - L_{n-1}^k) \, | \leqslant 2\Delta \, \} \, , \tag{60}$$

and $\mathfrak{L}_n^k$ denotes a set of the form

$$\mathfrak{L}_n^k = \mathfrak{L}_{n-1} \cap \tilde{\mathfrak{L}}_n^k \, , \quad n = 1, 2, \dots \, . \tag{61}$$

Geometric meaning of sets $\tilde{\mathfrak{L}}_n^k$ and $\mathfrak{L}_n^k$ is illustrated in **Fig. 9.** In this case, hyperstrip $\tilde{\mathfrak{L}}_n^k$ is obtained by uniting two different hyperstrips $\tilde{\mathfrak{L}}_n(y_n', U_n)$ and $\tilde{\mathfrak{L}}_n(y_n'', U_n)$ of form (7) with some y_n' and y_n'' such that one of the hyperplanes bounding a hyperstrip goes trough vertex L_{n-1}^k . Set $\mathfrak{L}_n^k$ is enclosed in heavily drawn lines.

Let us define the quantity

$$\delta^k(\mathfrak{L}_n^k) = \max_{L_n^j \in \mathfrak{L}_n^k} \{ \, | \, L_{n-1}^k - L_n^j \, | \, \} \, , \tag{62}$$

where L_n^j are the vertices of sets $\mathfrak{L}_n^k$.

Quantity $\delta^k(\cdot)$ is a certain pseudodiameter of set $\mathfrak{L}_n^k$ for the case when the maximum distance is sought only with respect to the isolated vertex L_{n-1}^k (which is a vertex of polyhedron $\mathfrak{L}_{n-1}$ and a vertex of set $\mathfrak{L}_n^k$) and therefore its calculation is easier than the calculation of true diameter $\mathfrak{L}_n^k$ according to (57). True is the following

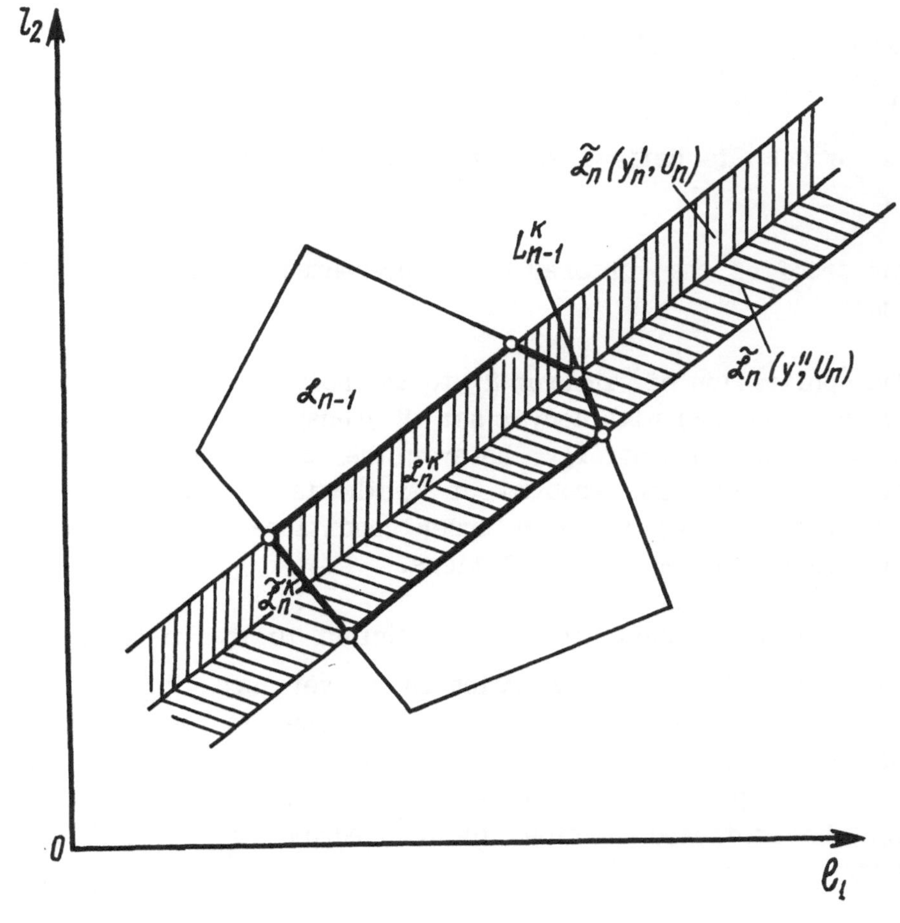

Fig. 9. Constructions of sets $\mathcal{L}_r^{(k)}$ and $\tilde{\mathcal{L}}_r^{(k)}$

Theorem 2. Solution of problem (59) coincides with the solution of problem

$$\min_{U_n \in \Pi} \max_{k \in \overline{1, N_{n-1}}} \{ \delta^k(\wp_n^k) \}, \tag{63}$$

where $\delta^k(\cdot)$ is defined according to (62) with regard to (60) and (61).

The proof of the **Theorem 2** is presented in [25] and it is left out here.

The application of **Theorem 2** to the solution of problem (56) (and therefore also (53) with convex set Ω) enables the computing efforts to be reduced substantially, however they remain to be rather high. Because of this, the problem of synthesis of simpler but fairly efficient algorithms is vital under conditions of severe constrains imposed on the volume of calculations.

Let diameter of estimate $\wp_{n-1}$ a priori to the n-th control step be known as well as the coordinates of vertices $\overset{*}{L}{}^j_{n-1}$ and $\overset{*}{L}{}^k_{n-1}$ defining this diameter according to (62). Then it seems to be natural to select suboptimal control $U_n = \bar{U}_n$ in such a way that the restricting hyperplanes in (7) are perpendicular to the line passing trough the mentioned vertices of polyhedron $\wp_{n-1}$. In this case, direction vector

$$\overset{*}{L}_{n-1} = \overset{*}{L}{}^j_{n-1} - \overset{*}{L}{}^k_{n-1} \tag{64}$$

and suboptimal control vector $\bar{U}_n$ are collinear. Then we obtain

$$\bar{U}_n = \overset{*}{\gamma}_n \overset{*}{L}_{n-1} , \tag{65}$$

where

$$\overset{*}{\gamma}_n = \arg \max_{\gamma_n : \gamma_n \overset{*}{L}_{n-1} \in \Pi} \{ | \gamma_n | \} . \tag{66}$$

The choice at each control step (64)-(66) is reasonably simple from computational point of view and in this case the main computing resources are spent for the realization of the operation of intersection of sets (7).

Let us now study the rate of convergence of algorithm (64)-(66). Designate by Ω_s the symmetrization of set Ω in the form

$$\Omega_s = \{ U \mid U \in \Omega \quad \text{or} \quad -U \in \Omega \} , \tag{67}$$

and the bound of the obtained set by $\bar{\Omega}_s$. Let us introduce the quantity

$$r = \inf_{U \in \bar{\Omega}_s} \{ \| U \| \} , \tag{68}$$

i.e. the value of the radius of the largest sphere with the center in the origin of the coordinates which belongs to Ω_s . Then for $r > 0$ in realizing the algorithm (64)-(66), the maximum possible thickness of the hyperstrip of form (7) is

$$\overset{*}{\rho} = 2\Delta / r .$$

In this case, true is the following

<u>Theorem 3.</u> If $r > 0$ and control U_n is selected according to algorithm (64)-(66) then with any realization of disturbances f_n satisfying (3), (4) the iterative procedure (8) of the refinement of a priori estimate (2) of 1-dimensional parameter vector L (using the results of measurements of scalar output of plant (1)) for each $\rho > \overset{*}{\rho}$, guarantees in a final number of steps N the achievement of a set with diameter

$$\delta(\Omega_n) \leqslant \bar{\delta} .$$

When this takes place, for the quantity $N = N(\vartheta, 1)$, where

$$\vartheta = \arccos(\overset{*}{\delta} / \bar{\delta}) , \tag{69}$$

which characterizes the rate of convergence of procedure (8), the following estimates are true:

$$N(\vartheta,3) < \frac{1}{1-\cos(\vartheta/2)} \, \frac{\pi}{2\sqrt{3}} ,$$

$$N(\vartheta,m) < \frac{\alpha(\pi/2,\, m-2)}{\alpha(\vartheta/2,\, m-2)} , \qquad m \neq 3 , \tag{70}$$

where

$$\alpha(\psi,k) = \int_0^{\psi} \sin^k\varphi \, d\varphi , \tag{71}$$

$$\alpha(\pi/2,m-2) = \frac{\Gamma[\,(m-1)/2\,]}{\Gamma(m/2)} \, \frac{\sqrt{\pi}}{2} . \tag{72}$$

The proof of **Theorem 3** is given in [25] and here it is omitted.

Let us note that the value of integral (71) can be determined analytically in the form

$$\alpha(\psi,k) = \begin{cases} \dfrac{(k-1)!!}{k!!}\,\psi - \cos\psi \displaystyle\sum_{i=0}^{k/2-1} \frac{(2i)!!}{k!!}\,\frac{(k-1)!!}{(2i+1)!!}\,\sin^{2i+1}\psi & \text{with even } k , \\[2ex] \dfrac{(k-1)!!}{k!!}\,(1-\cos\psi) - \cos\psi \displaystyle\sum_{i=1}^{(k-1)/2} \frac{(2i-1)!!}{k!!}\,\frac{(k-1)!!}{(2i)!!}\,\sin^{2i}\psi & \text{with odd } k . \end{cases}$$

The following designations are used here

$$k!! = k(k-2)(k-4)\ldots = \begin{cases} j!\,2^j & \text{with } k = 2j , \\ (2j)!/j!\,2^j & \text{with } k = 2j-1 , \end{cases}$$

$\Gamma(\cdot)$ is gamma-function.

As it follows from the content of the theorem, the suboptimal control (64)-(66) guarantees in asymptotics the obtaining of the limit estimate of the parameter vector in the form of a set with diameter not greater than the value δ^*. Moreover, the upper bound estimates are determined for a final number of control steps required to obtain sets with preset diameter $\bar{\delta} > \delta^*$.

Let us consider estimate (70) for $l = 2$. For example, let $r = 1$ and $\bar{\delta} = 2\sqrt{2}\,\Delta$. Then $\vartheta = \pi/4$, $N(\vartheta,2) < \pi/\vartheta = 4$, i.e. it is guaranteed that in 3 steps diameter δ^- is attained which is only $\sqrt{2}$ times greater than the maximum guaranteed diameter δ^*. Let us note that optimal control (with mutually perpendicular control vectors) guarantees achievement of the same diameter in 2 steps.

The active identification procedure with suboptimal control algorithm (64)-(66) described above was simulated in a computer for the case $l = 3$. The a priori estimate of unknown parameters l_i, $i=1, 2, 3$ was accepted in the form of a convex polyhedron with five vertices in three-dimensional space with diameter $\delta_0 = 1.71$.

Disturbance f_n was simulated by means of a standard subroutine for generation of random numbers uniformly distributed within interval [-0.2; 0.2]. Set Ω was set as a sphere with the unit radius and with the center in the origin of coordinates.

An alternative algorithm for solving the same problem was considered for comparison, with random uniform choice of control from set $\bar{\Omega}$, i.e. on the surface of the sphere.

The first control step with 100 different realizations of noise was carried out by means of both suboptimal and alternative algorithms. When this took place, the suboptimal algorithm has given better results in 97 cases, i.e. it has enabled the obtaining of sets with smaller diameter. Arithmetic mean diameter $\bar{\delta}_1$ of suboptimal algorithm has proved to be equal to 0.8756 and that of the alternative algorithm 1.3226, i.e. one and half time worse than the suboptimal one.

In addition, a multistep simulation was carried out with the results presented in **Fig. 10** . Solid line shows here the results of the suboptimal algorithm work and dotted line those of the alternative one. Presented are graphs of changes in diameters of sets Ω_n and in the number of their vertices on each iteration N_n .

Along with the active identification problem statement considered above, a different problem statement is possible. Namely: let N steps be assigned for studying the plant being investigated. It is required to plan the set of experiments for refining the parameter vector estimates, i.e. to find such succession of controls U_1, U_2, ..., U_N in such a way as to minimize diameter δ_N of set Ω_n in the last N-th step. In other words, it is required to find the solution of the original minimax "terminal" identification problem where unlike the "current" identification, important is only the final results of its solution. The "terminal" identification problem (which is much more difficult from the computational point of view, than the "current" identification problem) will be considered below in **Chapter 2** .

1.2.Guaranteed Estimetes of Linear Dynamic Systems Parameters

Let us extend the above method of set identification to the class of linear dynamic systems. In the majority of cases, their initial (primary) description is given in the differential equations language

$$\dot{X} = \tilde{\phi}(X,U,L,F) , \qquad (73)$$

where X, U, L and F are, respectively, state, control parameters and non-controllable disturbance vectors and $\phi(\cdot)$ is the given vector function.

However, since the identification and control algorithms studied below are oriented to computer realization, it is suitable to go over immediately from this description to description of plants motion (73) only in discrete instants by means of difference equations

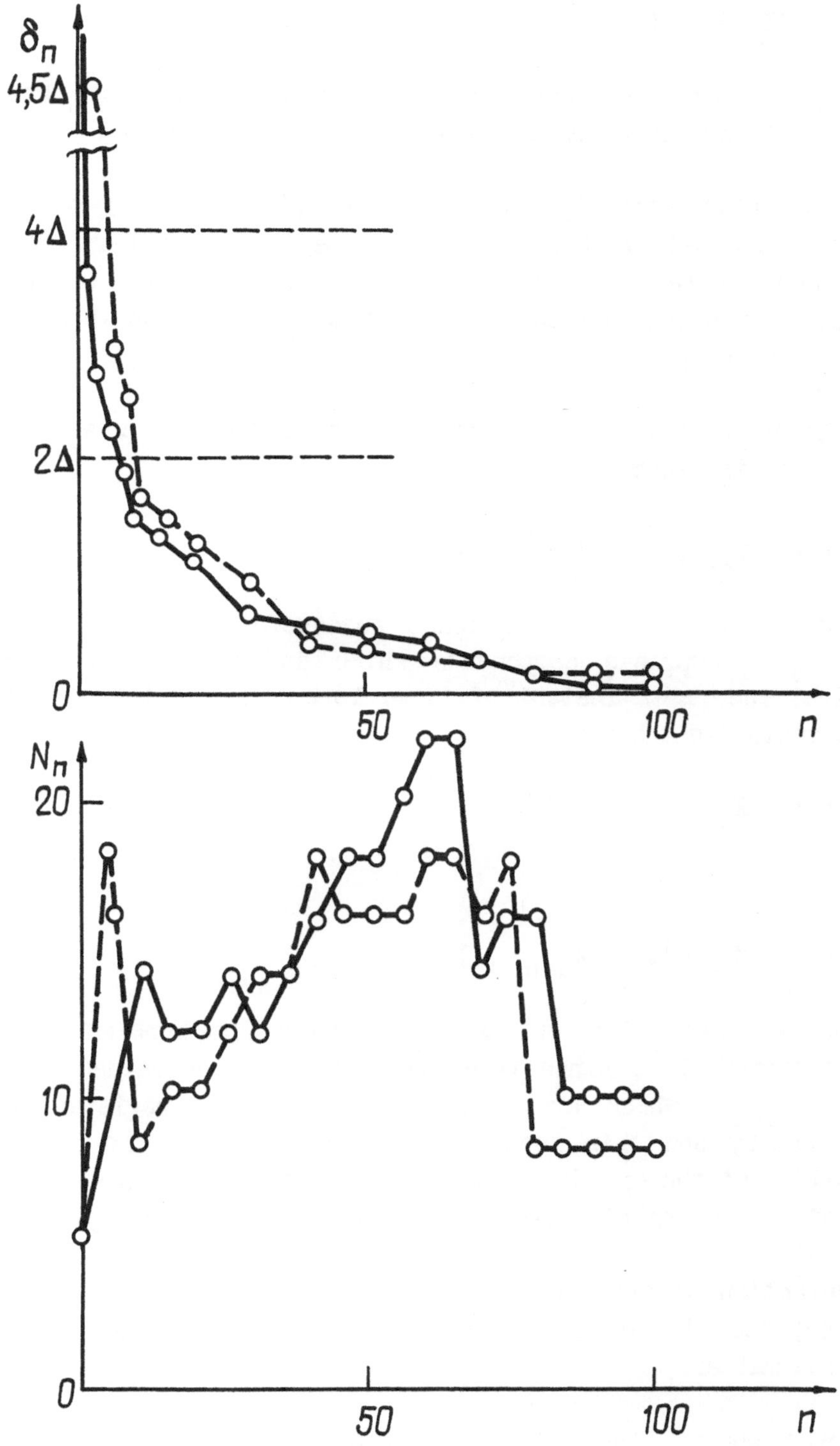

Fig. 10. Changing diameter of set $\mathfrak{L}_n$ and number of its vertices

$$X_{n+1} = \phi(X_n, U_n, L, F_n) \ , \quad n = 0,\ 1,\ 2,\ \dots , \tag{74}$$

where all designations have the same meaning as in (73), but with fixation of these variables at respective discrete instants of time.

The methods of obtaining difference equations are fairly well known and widely covered in the literature. Let us note here only one circumstance. There is a certain freedom of choice of the kind of phase space in which the motion of the system is described. Let us consider three main methods of them.

The choice of the still widely used state vector for a system of m-th order in the form

$$X_n^T = (\ X_n,\ X_{n+1},\ \dots,\ X_{n+(m-1)})$$

should be admitted to be the least successful since in this case vector X_n comprises components which in principle can not be measured at the n-th instant of time. More convenient is the choice of state vector in the form

$$X_n^T = (\ X_n,\ \dot{X}_n,\ \dots,\ X_n^{(m-1)}) \tag{75}$$

or

$$X_n^T = (\ X_n,\ X_{n-1},\ \dots,\ X_{n-(m-1)})\ . \tag{76}$$

In the last case, for $n > m - 1$ there is no problem at all in measuring vector X_n , since it consists only of the values of output variable x_{n-k} which are stored in respective register of the computer memory and this fact automatically removes the problem of observability of the dynamic system which makes it possible to abandon the use of Lueneberger's observers and its discrete analogs.

The selection of state vector X_n of two its possible forms (75) or (76) depends also on a number of factors, but we shall not dwell here on the matter.

Only a class of linear discrete systems with scalar control u_n

will be considered below in this chapter, for which equation (74) has the form

$$X_{n+1} = AX_n + Bu_n + Cf_n , \quad n = 0, 1, 2, \dots , \tag{77}$$

where A is matrix $(m \times m)$, B and C are m-dimensional vectors, f_n is a non-controllable scalar disturbance for which its a priori estimates (3) is given.

When true values of elements of matrix A and vectors B and C are unknown and only some rather rough a priori estimates are given for them to be refined in the future by means of one or other identification procedure, then, naturally, such a form of mathematical model should be selected from the class of models (77) for which the number of parameters to be identified is minimum. This requirement is just satisfied by the canonical form of equation (77) for which

$$A = \left\| \begin{array}{c|c} 0 & I_{m-1} \\ \hline \multicolumn{2}{c}{A_m^T} \end{array} \right\| , \qquad B = \left\| \begin{array}{c} 0 \\ \vdots \\ 0 \\ b \end{array} \right\| , \qquad C = \left\| \begin{array}{c} 0 \\ \vdots \\ 0 \\ 1 \end{array} \right\| , \tag{78}$$

where I_{m-1} is unitary matrix $(m-1) \times (m-1)$; A_m^T is the m-th row of matrix A .

In this case, parameter vector of system (77), (78)

$$L^T = (A_m^T, b) \tag{79}$$

has a minimum dimension equal to $(m+1)$. It is this from that we shall predominantly use below.

Let us note that in some cases the initial description of the equation of motion of the discrete system being studied is a scalar difference equation of the m-th order which we shall write in operator from for brevity sake:

$$H(D)x_n = gu_n + \bar{f}_n , \quad n = 0, 1, 2, \dots , \tag{80}$$

where

$$Dx_n = x_{n-1} , \qquad H(\cdot) = \sum_{i=0}^{m} h_i D^i .$$

Let us select state vector $\bar{X}_n$ of the system as

$$\bar{X}_n = (x_{n-1}, x_{n-2}, \ldots, x_{n-m}) , \tag{81}$$

then we obtain equation of its motion based on (80) and (81) in form

$$\bar{X}_{n+1} = \bar{A}\bar{X}_n + \bar{B}u_n + \bar{C}f_n , \tag{82}$$

where matrix $\bar{A}$ and vectors $\bar{B}$ and $\bar{C}$ have other canonical form:

$$\bar{A} = \left\| \begin{array}{c|c|c|c} -\frac{h_1}{h_0} & -\frac{h_2}{h_0} & \cdots & -\frac{h_m}{h_0} \\ \hline \multicolumn{3}{c|}{I_{m-1}} & 0 \end{array} \right\| , \quad \bar{B} = \left\| \begin{array}{c} \bar{b} \\ 0 \\ \vdots \\ 0 \end{array} \right\| , \quad \bar{C} = \left\| \begin{array}{c} 1 \\ 0 \\ \vdots \\ 0 \end{array} \right\| , \tag{83}$$

$$\bar{b} = \frac{g}{h_0} , \qquad f_n = \frac{1}{h_0} \bar{f}_n . \tag{84}$$

In this case, parameter vector L of the system has the form

$$L^T = (\bar{h}_1, \bar{h}_2, \ldots, \bar{h}_m, \bar{b}) , \text{ where } \bar{h}_i = -\frac{h_i}{h_0} , \quad i \in \overline{1,m} .$$

Below in **Chapter 2** , we shall consider also a more general case when scalar difference equation (80) has the form

$$H(D)x_n = G(D)u_n + f_n ,$$

where

$$G(\cdot) = \sum_{j=0}^{k} g_k D^k , \quad k < m$$

and it will be shown that also in this case the vector-matrix equation of this system can be written in canonical form.

Let us assume, that there is a priori estimate (2) for parameter vector L which should be refined by means of identification procedure. We shall consider first that simpler case when state vector X_n is measured without any noise.

Let there was estimated $L \in \Omega_n$ for vector L at the n-th step. Then at the $(n+1)$-th step, after measurement of values X_{n+1} and u_n, from system of equations (77), (78) or more exactly, from its last m-th equation we obtain that

$$L \in \tilde{\Omega}_{n+1} = \{ L \mid X_n^T \bar{L} + l u_n + f_n - x_{m,n+1} = 0 \} , \tag{85}$$

where $x_{m,n+1}$ is the m-th component of vector X_{n+1}, $\bar{L} = A_m$, $l = b$, $L^T = (\bar{L}^T, l)$.

Then, using recurrent procedure (8), we obtain finally the a posteriori estimate L in the form

$$L \in \Omega_{n+1} = \tilde{\Omega}_{n+1} \cap \Omega_n .$$

Thus, the problem of parametric identification of discrete system (77) accurate to designations is reduced to the identification problem already considered in detail above. As can be seen from (85), set $\tilde{\Omega}_{n+1}$ is a "hyperstrip" in parameter space $\{L\}$ also in the case being considered and, therefore, the sequence of estimates Ω_n is a sequence of convex polyhedra.

1.3. Guaranteed Estimates of Parameters of Linear Nonstationary Discrete Systems

Let us consider now the problem of identification of nonstationary systems with parameters (all or only some part of them) varying in time according to arbitrary laws. Like in the foregoing, we shall assume that these plants are subject to the action of non-controllable, limited in some sense, external disturbances. In this case the same as in the foregoing, mathematical models of the controlled plants will be used, however with the difference that they will explicitly take into account the non-stationarity of their parameters. Special features of the solution of the problem of nonstationarity parameter identification (in this case we can speak in fact about the tracking of the varying parameters) we shall consider following papers [26], [10] and using as example a class of nonstationary discrete systems with equation of motion taken in the form

$$X_{n+1} = A(L_n)X_n + B(L_n)u_n + Cf_n , \qquad n = 0, 1, 2, \dots , \tag{86}$$

where

$$A(\cdot) = \left\| \begin{array}{c|c} 0 & I_{m-1} \\ \hline \multicolumn{2}{c}{\bar{L}_n^T} \end{array} \right\| , \qquad B(\cdot) = \left\| \begin{matrix} 0 \\ \cdot \\ \vdots \\ \cdot \\ b_n \end{matrix} \right\| , \qquad C = \left\| \begin{matrix} 0 \\ \cdot \\ \vdots \\ \cdot \\ 0 \\ 1 \end{matrix} \right\| . \tag{87}$$

Here

$$L_n^T = (\bar{L}_n^T, b_n) \tag{88}$$

is (m+1)-dimensional vector of parameters varying in time in the general case, and the rest of designations has the same meaning as in the foregoing.

As to disturbance f_n , let us assume that a priori estimate (3), (4) remains valid for it.

The initial value L_0 of vector L_n is unknown and only its a priori estimate is given

$$L_0 \in \mathcal{Q}_0 , \tag{89}$$

where $\mathcal{Q}_0$ is a bounded set.

Since we can not expect to obtain constructive results in solving identification problem under so general assumptions concerning vector L_n, let us narrow the class of the plants being studied and hereinafter we shall restrict ourselves to the consideration of only such their sub-class for which the rate of change of vector L_n $\forall n \geq 0$ is restricted and these constraints are known to the investigator. Thus, let us assume that

$$\|\Delta L_n\| \leq \Delta_L = \text{const} \qquad \forall n \geq 0 , \tag{90}$$

where

$$\Delta L_n = L_{n+1} - L_n . \tag{91}$$

In this case, constraints are imposed on each component in the form of inequality

$$|\Delta l_{in}| \leq \delta_i = \text{const} , \quad i \in \overline{1,m+1} . \tag{92}$$

The constraints determine in the aggregate set $\delta\Omega$ and therefore the estimate of vector ΔL_n in the form

$$\Delta L_n \in \delta\Omega \qquad \forall n \geq 0 . \tag{93}$$

In solving problems of synthesis of control for nonstationary plants of the selected class, the problem of parameter identification becomes especially acute since the use only of a priori estimates of the parameters appears to be knowingly insufficient. In fact, the estimates of parameter vector constructed only on the basis of a priori estimates (89) and (93) prove to divergent over a time interval of any significant length. This can be best shown for the special case of set Ω_0 when this set represents a $(m+1)$-dimensional parallelepiped, i.e. when constraints are imposed upon the components of vector l_{i0} in the form

$$\underline{l}_i \leq l_{i0} \leq \bar{l}_i , \quad i \in \overline{1,m+1} , \tag{94}$$

where $\underline{l}_i$ and $\bar{l}_i$ are the set constraints.

Then we obtain from (91) and (93) estimates for components of vector L_n

$$\underline{l}_i - n\delta_i \leqslant l_{in} \leqslant \bar{l}_i + n\delta_i , \tag{95}$$

which in the limit at $n \to \infty$ give

$$- \infty \leqslant l_{i\infty} \leqslant \infty .$$

Therefore, to obtain consistent estimates of the vector, it is necessary to use a posteriori estimates obtained from the results of the measurements taken along the trajectory of the motion of the plant being studied.

Let us note, that in identification of a nonstationary plant, there exists an unavoidable in principle delay for one step of the discrete system operation. Indeed, it can be seen from equation (80) that in measuring at the $(n+1)$-th instant the value of vector X_{n+1}, this equation determines some set $\tilde{\wp}_n^{n+1}$, defining a current a posteriori estimate of the previous value of the parameter vector, i.e. of vector L_n, which jointly with the previous a priori estimate for the n-th instant, i.e. with the estimate

$$L_n \in \wp_n^n , \tag{96}$$

determines the a posteriori estimate of the vector L_n

$$L_n \in \wp_n^{n+1} = \tilde{\wp}_n^{n+1} \cap \wp_n^n . \tag{97}$$

By virtue of the fact that for the solution of one or other problem of control at the $(n+1)$-th instant of time, the knowledge of vector L_{n+1} (more exactly, of its estimate) at the same instant of time is required then this generates a need for solving the extrapolation (forecasting) problem. It is required, using available estimate (97), to construct in one way or another at the $(n+1)$-th instant of time the estimate of vector L_{n+1}, i.e. to obtain estimate

$$L_{n+1} \in \wp_{n+1}^{n+1} . \tag{98}$$

It is obvious that the only base for solution of this problem is the availability of equation (91) and estimate (93). Since vectors L_n and ΔL_n are specified accurate to their belonging to some sets, then it is obvious that also for vector L_{n+1} we can obtain only an estimate of the form

$$L_{n+1} \in \mathfrak{L}^{n+1}_{n+1} = \mathfrak{L}^{n+1}_{n} + \delta\mathfrak{L} . \tag{99}$$

The sum of independent sets in this expression is understood as the Minkowski's sum, i.e.

$$\mathfrak{Z} = \mathfrak{X} + \mathfrak{Y} = \bigcup_{\substack{X \in \mathfrak{X} \\ Y \in \mathfrak{Y}}} (X + Y) . \tag{100}$$

At first, in an effort to simplify the solution of the parametric identification problem, we shall consider the special case when $f_n = \emptyset$ $\forall\, n \geqslant 0$. Let the controlled plant (86) be changed over from state X_0 to state X_1 under the effect of control selected in some way. Then from (86) with $f_n \equiv 0$ we obtain

$$x_{m,1} = \bar{L}^T_0 X_0 + b_0 u_0 . \tag{101}$$

This equation determines hypersurface in space $\{L\}$ defining set $\tilde{\mathfrak{L}}^1_0$ such that

$$L_0 \in \tilde{\mathfrak{L}}^1_0 . \tag{102}$$

From (89) and (102) we obtain a posteriori estimate in the form

$$L_0 \in \mathfrak{L}^1_0 = \tilde{\mathfrak{L}}^1_0 \cap \mathfrak{L}^0_0 . \tag{103}$$

Next, it is required to construct the estimate for vector L_1 , i.e. to solve the forecasting problem, with $n = 1$ and using estimate (103) (which is always better than estimate (89) by definition). The basis for solution of this problem is condition (90) from which we obtain that $L_1 = L_0 + \Delta L_0$. Since vectors L_0 and δL_0 are given accurate only to their belonging to sets $\mathfrak{L}_0$ and $\delta\mathfrak{L}$, respectively, for vector L_1 we obtain also estimate

$$L_1 \in \mathfrak{L}_1^1 = \mathfrak{L}_0^1 + \delta\mathfrak{L} , \tag{104}$$

where, as mentioned earlier, the sum of independent sets is understood as the Minkowski's sum.

After change-over of the controlled plant from state X_1 to state X_2 under the effect of control u_1 , we obtain from (86)

$$x_{m,2} = \bar{L}_1^T X_1 + b_1 u_1 . \tag{105}$$

This equation determines set $\tilde{\mathfrak{L}}_1^2$. From equation (105) we obtain by analogy with (103) the a posteriori estimate for L_1

$$L_1 \in \mathfrak{L}_1^2 = \tilde{\mathfrak{L}}_1^2 \cap \mathfrak{L}_1^1 . \tag{106}$$

Next, similarly to the foregoing, the estimate of the forecasted value of L_2 is constructed

$$L_2 \in \mathfrak{L}_2^2 = \mathfrak{L}_2^1 + \delta\mathfrak{L} . \tag{107}$$

Then constructions of the form (106) and (105) are repeated for all next values of n .

It follows from the proposed procedure for constructing sets $\mathfrak{L}_n^n$ that in this case both the operation of summarion of sets is used which results in their "increase" and the intersection operation "decreasing" them. Because of this, it follows from intuitive considerations that the estimates of the form (106) and (107), as distinct from estimates (99) constructed only on the basis of a priori estimates (93) and (89), can not be divergent with $n \to \infty$. Nevertheless, let us give a rigorous proof of the verity of this statement.

Let us show the validity of the

Theorem 4. Let for parameter vector of discrete system (86) satisfying difference equation

$$L_n = L_{n-1} + \Delta L_{n-1} ,$$

where vector ΔL_{n-1} is bounded by estimate (90), the a priori estimate be set at the (n-1)-th step:

$$L_{n-1} \in \wp_{n-1}^{n-1} . \tag{108}$$

Then the recurrent procedure

$$L_{n-1} \in \tilde{\wp}_{n-1}^{n} , \tag{109}$$

where

$$\tilde{\wp}_{n-1}^{n} = \{ L_{n-1} \mid \bar{L}_{n-1}^{T} X_{n-1} + b_{n-1} u_{n-1} - x_{m,n} = 0 \} , \tag{110}$$

$$L_{n-1} \in \wp_{n-1}^{n} = \tilde{\wp}_{n-1}^{n} \cap \wp_{n-1}^{n-1} , \tag{111}$$

$$L_n \in \wp_n^n = \wp_{n-1}^{n} + \delta\wp , \tag{112}$$

with linearly independent vectors $Z_{n-1}^{T} = (X_{n-1}^{T}, u_{n-1})$ determines sequence of bounded sets $\wp_n^n$, i.e. $\delta_n = \delta(\wp_n^n) < \infty$ with $n \to \infty$.

Let us present the proof of the theorem. It was shown above that recurrent procedures (110)-(112) are equivalent to the solution of the respective system of linear algebraic equations with indeterminate form in its right-hand side. This statement is true also for the solution of the system of linear equations under conditions of **Theorem 4**, based on successive elimination of variables being sought. Really, with $n \geq (m+1)$ there is a system of equations of the form (101), i.e. of estimates

$$x_{m,n+1} = \bar{L}_n^T X_n + b_n u_n$$

and, respectively,

$$x_{m,n+1-k} = X_n^T(\bar{L}_n - \sum_{j=1}^{k} \Delta\bar{L}_{n-1-j}) + (b_n - \sum_{j=1}^{k} \Delta b_{n-1-j})u_n , \quad k<n . \tag{113}$$

Assuming in (113) $k = 1, 2, \ldots, m+1$, we obtain the following system of equations

$$x_{m,n} = X^T_{n-1}\bar{L}^T_{n-1} + b_{n-1}u_{n-1} ,$$

$$x_{m,n-1} = X^T_{n-2}(\bar{L}^T_{n-1} - \Delta\bar{L}^T_{n-2}) + (b_{n-1} - \Delta b_{n-2})u_{n-2} ,$$

$$\vdots$$

$$x_{m,n-m} = X^T_{n-m-1}(\bar{L}^T_{n-1} - \sum_{j=1}^{m} \Delta\bar{L}^T_{n-1-j}) + (b_{n-1} - \sum_{j=1}^{m} \Delta b_{n-1-j})u_{n-m-1} .$$

Denote

$$\hat{X}^T_n = (x_{m,n}, x_{m,n-1}, \ldots, x_{m,n-m-1}) ,$$

$$\mathfrak{Z}_{n-1} = \left\| \begin{matrix} Z^T_{n-1} \\ Z^T_{n-2} \\ \cdot \\ \cdot \\ \cdot \\ Z^T_{n-m-1} \end{matrix} \right\| , \qquad \mathfrak{D}_{n-2} = \left\| \begin{matrix} Z^T_{n-2}\Delta L_{n-2} \\ Z^T_{n-3}\Delta L_{n-3} \\ \cdot \\ \cdot \\ \cdot \\ Z^T_{n-m-1}\Delta L_{n-m-1} \end{matrix} \right\| ,$$

and we can obtain this system in the more compact form

$$\hat{X}_n = \overset{*}{Z}_{n-1}L_{n-1} - \mathfrak{D}_{n-2} . \tag{114}$$

If $\det \tilde{Z}_{n-1} \neq 0$, then we obtain from (114)

$$L_{n-1} = \tilde{Z}^{-1}_{n-1}\hat{X}_n + \tilde{Z}^{-1}_{n-1}\mathfrak{D}_{n-2} . \tag{115}$$

Since true values of vector ΔL_n are unknown and their values are given only up to their belonging to set $\delta\mathfrak{L}$, therefore set $\mathfrak{D}_{n-2}$ is not single-point one, but it is bounded (with $\det \tilde{Z}_{n-1} \neq 0$) by virtue of condition (93) accepted above. Thus, estimates Ω_n^n $\forall\, n \geq 0$ are restricted which proves the statement.

The obvious result follows from (115) that when $\| \Delta L_n \| \to 0$, i.e. when $L_n = L_0 = \overset{*}{L}$, then at $L_{n-1} = \overset{*}{L}$ $n = m+1$.

It follows from expressions (111) and (112) that sets Ω_n^n are convex since the class of convex polyhedra is closed not only with respect to the operation of intersection but also with respect to the operation of summation.

Let us dwell briefly on the execution of summation (99) of sets Ω_{n-1}^n and $\delta\Omega$ which was not encountered before. It follows from (92) and (93) that set $\delta\Omega$ is given in the form

$$\delta\Omega = \{ l_i \mid -\delta_i \leq l_i \leq \delta_i , \quad i \in \overline{1,m+1} \} .$$

Let us denote the vertices of polyhedron Ω_{n-1}^n as

$${}^kL_{n-1}^n = \left\| {}^kl_{i,n-1}^n \right\|_{i=1}^{m+1} , \quad k \in \overline{1,N_{n-1}} ,$$

where N_{n-1} is the number of vertices of set Ω_{n-1}^n. On execution of operation of summation of single-point set ${}^kL_{n-1}^n$ with set $\delta\Omega$ we obtain set

$${}^kL_n^n = \{ l_i \mid {}^kl_{i,n-1}^n - \delta_i \leq l_i \leq {}^kl_{i,n-1}^n + \delta_i , \quad i \in \overline{1,m+1} \} . \tag{116}$$

Then, in accordance with (112) and (116) set Ω_n^n is defined as the convex enveloping surface of the system of vectors ${}^kL_{i,n-1}^n$, i.e.

$$\Omega_n^n = \underset{k \in \overline{1,N_{n-1}}}{\mathrm{con}\, V} \{ {}^kL_{i,n-1}^n , \quad i \in \overline{1,m+1} \} . \tag{117}$$

Let us consider now the general case of parameters identification of system (86) with $f \neq \emptyset$. In this case, set $\tilde{\Omega}_{n-1}^n$ is defined instead of expression (110) by the following expression

$$\tilde{\Omega}_{n-1}^n = \{ L_{n-1} \mid X_{n-1}^T A_{n-1} + u_{n-1} b_{n-1} + f_n - x_{m,n} = 0 \} , \tag{118}$$

where $f_n \in \mathcal{f}$ in accordance with (3). All other stages of the identification procedure (100)-(112) remain unchanged.

1.4. Simultaneous Construction of Parameter Vectors and State Vectors Estimates of Linear Dynamic Systems

Control system designer rather often is found to be in situation when the a priori estimates of controlled plant parameters are specified very coarsely (approximately) and, therefore, they need refinement and the phase coordinate vector of the system is measured with intensive noise which already can not be neglected.

Let us consider this situation as applied to the class of dynamic systems (77) already analyzed above introducing to its description required changes and additions. Thus, assume that the motion of the dynamic system considered here is described as before by equation (77) with canonical structure of matrix A and vectors B and C, and assume that m-dimensional vector is measured instead of vector X_n:

$$Y_n = SX_n + Z_n , \tag{119}$$

where Z_n is m-dimensional noise vector, S is given non-singular matrix.

There is a practically countless number of papers in which, under the assumption that X_0, L are random vectors and Z_n and f_n are random vectors and scalar sequences, respectively, the problem is stated of constructing optimal in some sense estimates of vectors X_n and L using current measurements results. However, we shall consider here the problem of constructing guaranteed estimates of vectors X_n and L under the assumption that f_n is a bounded disturbance for which its a priori estimate (3), (4) is given, a priori estimate (2) is given for parameter vector L as before and we shall assume that a priori estimate of quantities X_0 and Z_n are given in form

$$X_0 \in \mathfrak{X}_0 , \tag{120}$$

$$Z_n \in \mathfrak{Z} \quad \forall\, n \geqslant 0 , \tag{121}$$

where $\mathfrak{X}_0$ and $\mathfrak{Z}$ are given bounded convex sets. It is obvious that in this case the well-known methods for solving the problem of construction of a sequence of estimates X_n and N based on its stochastic statement appear to be unsuitable.

First of all, let us define more accurately the nature of sets $\mathfrak{L}_0$ and $\mathfrak{X}_0$ given a priori. Let us assume sets $\mathfrak{X}_0$ and $\mathfrak{L}_0$ to be convex polyhedra in the spaces of respective dimensions R^{m+1} and R^m specified by the vertices-vectors L_0^i , $i \in \overline{1,N_0}$ and X_0^j , $j \in \overline{1,M_0}$ where N_0 and M_0 are the numbers of the vertices of polyhedra $\mathfrak{L}_0$ and $\mathfrak{X}_0$, respectively. Thus, let us conclusively assume that matrices of vertices are specified for $\mathfrak{L}_0$ and $\mathfrak{X}_0$

$$G_L^0 = \| L_0^1, L_0^2, \ldots, L_0^{N_0} \| , \tag{122}$$

$$G_X^0 = \| X_0^1, X_0^2, \ldots, X_0^{M_0} \| . \tag{123}$$

Let us assume $\mathfrak{Z}$ in form of a m-dimensional rectangle with center in the origin of coordinates and with sides equal to $\pm\Delta_{zi}$, $i \in \overline{1,m}$.

It is required to construct the sequence of estimates of vectors X_n and L from the results of measurements of values Y_n and u_n , $n = 0, 1, 2, \ldots$ and a priori estimates (2), (3), (4), (120) and (121). The general scheme of solution of the formulated problem was briefly outlined in paper [27] and then in some more detail in paper [28]. We shall follow this scheme hereinafter in describing the construction of sets $\mathfrak{X}_n$ and $\mathfrak{L}_n$ such that

$$L \in \mathfrak{L}_{n+1} \subseteq \mathfrak{L}_n , \tag{124}$$

$$X_{n+1} \in \mathfrak{X}_{n+1} \subseteq \tilde{\mathfrak{X}}_{n+1} , \tag{125}$$

where

$$\tilde{\mathfrak{X}}_{n+1} = S^{-1}Y_{n+1} - S^{-1}\mathfrak{Z} . \tag{126}$$

Here, the algebraic sum of sets is understood, like above, as their Minkowski's sum whereas their linear transformations and calculations of scalar products are carried out in accordance with the rows presented in [10].

Let at the n-th instant of time, estimates for vectors L and X_n be

$$L \in \mathfrak{L}_n , \tag{127}$$

$$X_n \in \mathfrak{X}_n , \tag{128}$$

where $\mathfrak{L}_n$ and $\mathfrak{X}_n$ are polyhedra for which their matrices G^n_L and G^n_X , respectively, are given. With $n = 0$, such polyhedra are given sets $\mathfrak{L}_0$ and $\mathfrak{X}_0$.

It follows from (77) and (127), (128), (3), (4) that at the (n+1)-th instant of time the guaranteed forecasted estimate of vector X_{n+1} has the form

$$X_{n+1} \in \bar{\mathfrak{X}}_{n+1} = \bigcup_{L \in \mathfrak{L}_n} [A(L)\mathfrak{X}_n + B(L)u_n + Cf] . \tag{129}$$

From two consistent estimates of vector X_n (129) and (126), we obtain finally its a posteriori estimate

$$X_{n+1} \in \mathfrak{X}_{n+1} = \tilde{\mathfrak{X}}_{n+1} \cap \bar{\mathfrak{X}}_{n+1} . \tag{130}$$

Let us now turn to determination of the estimate $\mathfrak{L}_{n+1}$ of vector L from the measurement results of the form (119) and estimates (127), (128) and (129). From the last equation of system (77), taking into account (78), we obtain the following estimate of the vector

$$L \in \tilde{\mathfrak{L}}_{n+1} = \{ L \mid \bigcup_{X_n \in \mathfrak{X}_n} (X_n^T \bar{L} + bu_n + f - \mathfrak{x}_{m,n+1} = 0 \} , \tag{131}$$

where $\mathfrak{x}_{m,n+1}$ is the projection of set $\mathfrak{X}_{n+1}$ on the axis x_m , i.e. the set which is determined as follows

$$\mathfrak{x}_{m,n+1} = \{ x_{m,n+1} \mid \underline{x}_{m,n+1} \leq x_{m,n+1} \leq \bar{x}_{m,n+1} \} , \tag{132}$$

where

$$\underline{x}_{m,n+1} = \inf_{k\in\overline{1,M_{N+1}}} \{ x^k_{m,n+1} \} , \tag{133}$$

$$\bar{x}_{m,n+1} = \sup_{k\in\overline{1,M_{N+1}}} \{ x^k_{m,n+1} \} , \tag{134}$$

Here $x^k_{m,n+1}$ is the m-th component of the k-th vertex X^k_{n+1} of polyhedron $\mathfrak{X}_{n+1}$ and M_{n+1} is the number of its vertices, in accordance with designations accepted above.

Then we obtain finally from two consistent estimates (127) and (131) the a posteriori estimate

$$L \in \mathfrak{L}_{n+1} = \tilde{\mathfrak{L}}_{n+1} \cap \mathfrak{L}_n . \tag{135}$$

The obtained system of equations (129)-(135) jointly with (127), (128) form a closed system of estimates which determines identically the evolution of sets $\mathfrak{L}_n$ and $\mathfrak{X}_n$ from the results of current measurements (119) and a priori estimates (2)-(4), (120), (121). However, the obtained system of equations is non-linear which makes its solution difficult. Because of this, let us describe a constructive method of its solution which takes account of structural features of this system.

First, let us seek for solutions of equations (129) and (130) enabling $\mathfrak{X}_{n+1}$ to be determined in the form

$$\mathfrak{X}_{n+1} = \mathfrak{F}_1(\mathfrak{X}_n, \mathfrak{L}_n, Y_n, \mathfrak{Z}, u_n, f) . \tag{136}$$

Its substitution into (133), or more exactly, its projection onto axis x_m and the following substitution into (131) makes it possible to find $\mathfrak{L}_{n+1}$ in the form

$$\mathfrak{L}_{n+1} = \mathfrak{F}_2(\mathfrak{L}_n, \mathfrak{X}_n, \mathfrak{x}_{m,n+1}, u_n, f) . \tag{137}$$

Now, let us dwell on the method of constructing set $\bar{\mathfrak{X}}_{n+1}$ defined by equation (129). It follows from equation (129) that since this transformation is linear, we can conclude that if $\mathfrak{X}_n$ is a convex polyhedron then $\bar{\mathfrak{X}}_{n+1}$ will be also a convex polyhedron. For the subsequent construction of set $\bar{\mathfrak{X}}_{n+1}$, let us write expression (129) in the following form, taking into account the structure of matrix

$$\bar{\mathfrak{X}}_{n+1} = \mathfrak{F}(\mathfrak{X}_n, \mathfrak{L}_n, u_n, f) = \left\| \frac{\bar{A}\mathfrak{X}_n}{\bigcup\limits_{X_n \in \mathfrak{X}_n, L \in \mathfrak{L}_n} (X_n^T \bar{L} + bu_n + f)} \right\|, \tag{138}$$

where $\bar{A}$ is a numerical matrix obtained from matrix A by deleting its last row. It follows from (138) that since set $\mathfrak{L}_n$ is not a single-point one, each point X_n of set $\mathfrak{X}_n$ is transformed to some subset (along axis x_m) in set $\bar{\mathfrak{X}}_{n+1}$ which naturally, makes its construction more difficult. Because of this, to construct $\bar{\mathfrak{X}}_{n+1}$, let us use some generalization of the corollary to lemma presented in [26], namely, we shall use the following

<u>Statement 2.</u> If X_i and Y_j are vertices of arbitrary polyhedra $\mathfrak{X}$ and $\mathfrak{Y}$, respectively, then

$$\sup_{\substack{X \in \mathfrak{X} \\ Y \in \mathfrak{Y}}} \{ v = X^T Y \} = \sup_{\substack{i \in \overline{1,N} \\ j \in \overline{1,M}}} \{ \tilde{v}_{ij} = \tilde{X}_i^T \tilde{Y}_j \} \tag{139}$$

and

$$\inf_{\substack{X \in \mathfrak{X} \\ Y \in \mathfrak{Y}}} \{ v = X^T Y \} = \inf_{\substack{i \in \overline{1,N} \\ j \in \overline{1,M}}} \{ \tilde{v}_{ij} = \tilde{X}_i^T \tilde{Y}_j \}, \tag{140}$$

where $\tilde{X}_i$ and $\tilde{Y}_j$ are vertices of convex envelopes $\tilde{\mathfrak{X}}$ and $\tilde{\mathfrak{Y}}$ of sets $\mathfrak{X}$ and $\mathfrak{Y}$, respectively, and N and M are the numbers of these vertices.

It follows from (138) and (139), (140) that it is sufficient for constructing set $\bar{\mathfrak{X}}_{n+1}$ to subject to transformation (138) each of the vertices X_n^i, $i \in \overline{1,M_n}$ where M_n is the number of vertices of polyhedron $\mathfrak{X}_n$ and the convex envelope of the vectors obtained in this way will be the set $\bar{\mathfrak{X}}_{n+1}$.

Let us introduce the following designations

$$\bar{\delta}^i_{n+1} = \sup_{j \in \overline{1,\bar{N}_n}} \{ \bar{L}_j^T X_n^i \} , \tag{141}$$

$$\underline{\delta}^i_{n+1} = \inf_{j \in \overline{1,\bar{N}_n}} \{ \bar{L}_j^T X_n^i \} , \tag{142}$$

where $\bar{L}_j$ is the j-th vertex of polyhedron $\bar{\mathfrak{L}}_n$ which is the projection of set $\mathfrak{L}_n$ onto subset $\{L\}$ and $\bar{N}_n$ is the number of vertices of polyhedron $\bar{\mathfrak{L}}_n$.

$$\bar{x}^i_{m,n+1} = \bar{\delta}^i_{n+1} + \Delta_f + \left| \begin{array}{ll} \bar{b}u_n , & u_n > 0 , \\ \underline{b}u_n , & u_n < 0 , \end{array} \right\} \tag{143}$$

$$\underline{x}^i_{m,n+1} = \underline{\delta}^i_{n+1} - \Delta_f + \left| \begin{array}{ll} \bar{b}u_n , & u_n < 0 , \\ \underline{b}u_n , & u_n > 0 , \end{array} \right\} \tag{144}$$

where

$$\bar{b} = \sup_{b \in \mathfrak{b}_n} \{ b \} , \qquad \underline{b} = \inf_{b \in \mathfrak{b}_n} \{ b \} .$$

Here $\mathfrak{b}_n$ is the projection of set $\mathfrak{L}_n$ onto axis b, defined similarly to the projection of set $\mathfrak{X}_{n+1}$ onto axis x_m (ref. to (132)-(134)).

It follows from (138), as well as from statement (2) and (141)-(144) that each of vectors X_n^i generates some subset whose border elements are determined by the following pair of vectors

$$\bar{X}_{n+1}^i = \left\| \frac{\bar{A}X_n^i}{\bar{X}_{m,n+1}^i} \right\| , \qquad (145)$$

$$\underline{X}_{n+1}^i = \left\| \frac{\bar{A}X_n^i}{\underline{X}_{m,n+1}^i} \right\| \qquad (146)$$

and, therefore, we obtain finally that

$$\bar{\mathfrak{X}}_{n+1} = \underset{i \in \overline{1,M_n}}{\mathrm{conv}} \{ \bar{X}_{n+1}^i , \underline{X}_{n+1}^i \} . \qquad (147)$$

Generally, the operation of determination of the convex envelope of an arbitrary system of vectors is rather complex problem, but its execution in the present case is essentially simplified because $\mathfrak{X}_n$ is a complex polyhedron. For a more detailed description of the algorithm of constructing set $\bar{\mathfrak{X}}_{n+1}$ from expressions (145),(146) and (141)-(144) ref. to [28].

With the measured value Y_{n+1}, the estimate for X_{n+1} follows from (119)and (121)

$$X_{n+1} \in \tilde{\mathfrak{X}}_{n+1} = S^{-1}Y_{n+1} - S^{-1}\mathfrak{Z} . \qquad (148)$$

From two consistent estimates (148) and (129) defining X_{n+1} accurate to its membership in some sets, we obtain the following relationship determining the sequence of a posteriori estimates:

$$X_{n+1} \in \mathfrak{X}_{n+1} = \tilde{\mathfrak{X}}_{n+1} \cap \bar{\mathfrak{X}}_{n+1} , \quad n = 0, 1, 2, \dots . \qquad (149)$$

Because sets $\tilde{\mathfrak{X}}_{n+1}$ and $\bar{\mathfrak{X}}_{n+1}$ are convex polyhedra, their intersection generates also a convex polyhedron. The operation of their intersection was considered above. Now, let us show that for procedure (149) holds

Statement 3. The first (m-1) components of estimate $\mathfrak{X}_{n+1}$ defined by expressions (138), (148) and (149) with S = I satisfy conditions

$$\mathfrak{x}_{i,n+1} \subset \tilde{\mathfrak{x}}_{i,n+1} , \quad i \in \overline{1,m-1} , \tag{150}$$

i.e. the application of procedure (149) to the first (m-1) components of vector X_{n+1} makes it possible to obtain better estimates then estimate (148).

Let us prove the truth of this statement. Let us assume that the following relationships hold for estimate

$$\underline{x}_{in} \leqslant \overset{*}{x}_{in} \leqslant \bar{x}_{in} , \quad i \in \overline{1,m} , \tag{151}$$

where $\overset{*}{x}_{in}$ is the true value of the i-th component of vector X_n .

For the sake of simplicity, let us assume that

$$\Delta_{zi} = \Delta_z \quad \forall\ i \in \overline{1,m} . \tag{152}$$

Then from the properties of set $\mathfrak{Z}$ we obtain that

$$\bar{x}_{in} - \underline{x}_{in} = 2\Delta_z \quad \forall\ i \in \overline{1,m} . \tag{153}$$

Next, from (77) and (78) we obtain that

$$\overset{*}{x}_{i-1,n+1} = \overset{*}{x}_{in} , \quad i \in \overline{2,m} . \tag{154}$$

then from (148), (152) and (154) it follows that

$$y_{i-1,n+1} = \overset{*}{x}_{in} + z_{i,n+1} , \quad i \in \overline{2,m} .$$

Since the limiting values of $z_{i,n+1}$ are equal to $\pm\Delta_z$, we obtain the estimates satisfying condition (150) for each i-th component, $i \in \overline{2,m}$ of vectors X_{n+1} constituting set $\mathfrak{X}_{n+1}$, by executing the set intersection operation (149).

To answer the question, under which conditions the relationship of the form (150) holds also for the first component of the estimate $\mathfrak{X}_{n+1}$, requires this time to take into account simultaneously and more completely the specific features of sets $\mathfrak{L}_n$, $\mathfrak{X}_n$, $\mathfrak{X}_{n+1}$ and generally it can not be obtained in analytical form.

Now, once we have considered the method of constructing the estimate of state vector X_{n+1}, we can turn to the consideration of the second half of the set problem, namely to the construction of estimate $\mathfrak{L}_{n+1}$ from expression (135). First of all, let us note that set $\tilde{\mathfrak{L}}_{n+1}$ defined by expression (131) is nonconvex even with convex set $\mathfrak{X}_n$ which not only makes it difficult to construct the bounds of this set, but, what is especially important, makes it essentially difficult to perform the operation of its intersection with set $\mathfrak{L}_n$. It is easy to show that a sequential execution of operation (135) is virtually a sequential solution of the system of linear (with respect to vector L) algebraic equations given uncertainty in coefficients of its both right-hand and left-hand sides, with the following projection of the obtained result onto initial a priori set $\mathfrak{L}_0$. The problem of obtaining solution of such system of equations in such a general statement does not possess a constructive solution. However, its constructive solution was already obtained in **Section 1.1.** for the case when set $\mathfrak{X}_n$ is a m-dimensional rectangle. To take advantage of these results, let us reduce this problem to the already solved one. With this aim in view, let us substitute set $\mathfrak{X}_n$ by m-dimensional rectangle $\overset{*}{\mathfrak{X}}_n \subseteq \mathfrak{X}_n$ circumscribed about the set. The construction of the rectangle with given vertices of set $\mathfrak{X}_n$ does not involve any difficulties. The replacement of set $\mathfrak{X}_n$ by set $\overset{*}{\mathfrak{X}}_n$ in (131) results respectively also in the replacement of estimate $\tilde{\mathfrak{L}}_{n+1}$ by the coarsened estimate $\check{\mathfrak{L}}_{n+1} \supseteq \tilde{\mathfrak{L}}_{n+1}$ calculated as in the following way

$$\check{\mathfrak{L}}_{n+1} = \{ L \mid \bigcup_{X_n \in \overset{*}{\mathfrak{X}}_n} (X_n^T \bar{L} + bu_n + f - x_{m,n+1} = 0 \} . \tag{155}$$

In this case, a corresponding substitution should be carried out also in (135), i.e. the procedure

$$L \in \mathfrak{L}_{n+1} = \check{\mathfrak{L}}_{n+1} \cap \mathfrak{L}_n \,, \quad n = 0, 1, 2, \dots \,. \tag{156}$$

is used in what follows instead of (135). The realization of this procedure with the stipulated properties of set $\overset{*}{\mathfrak{X}}_n$ is carried out in the way described in the foregoing.

It follows from the analysis of the above procedure of the simultaneous (at each n-th step) construction of the estimates of the parameter and state vectors that the availability of the uncertainty in the state vector estimate is "transferred" to additional uncertainty in the parameter vector estimate thus deteriorating them.

It is obvious that the "quality" of solution of these problems most essentially depends on the properties of the control sequence of the dynamic system involved. This important problem (the finding of the identifiability conditions under the conditions of the simultaneous construction of the state vector) is worthy of separate consideration. Let us note here only that when using the set identification procedure (156), a part of measurements (119) may prove to be noninformative in the sense that their use may not result to the improvement of estimate $\mathfrak{L}_{n+1}$ as compared with estimate $\mathfrak{L}_n$.

1.5. Guaranteed Estimates in Systems Nonlinear in Parameters

Only systems linear in parameters have been considered throughout in the foregoing in this **Chapter** . Now, let us study the solution of the problem of obtaining guaranteed estimates for systems nonlinear in parameters, considering a class of systems without memory. Let the plant being studied be described instead of (1) by the equation

$$y_n = \varphi(U_n, L) + f_n \,, \qquad n = 1, 2, \dots \,,$$

where $\varphi(\cdot)$ is a scalar function nonlinear at least in argument L .

Let us rewrite this equation in the form

$$\varphi(L,f_n,U_n) = y_n - \psi(U_n,L) - f_n = 0 , \qquad n \in \overline{1,N} . \tag{157}$$

Here, like above, f_n is the uncontrolled disturbance for which its a priori estimate is given. For the unknown vector L , its a priori estimate is given in form (2). It is required to propose a recurrent procedure for refining this estimate using equations (157) where quantities y_n and U_n are taken to be unknown.

Let us introduce the "canonical" designation of the vector of the variable sought for $L = X$ and then let us "imbed" the problem of finding the "solutions" of system (157) into the problem of finding the "solutions" of the following system

$$\varphi_i(X,S_i) = 0 , \qquad i \in \overline{1,n} , \tag{158}$$

where vector S_i is taken generally to mean the k-dimensional vector of unknown parameters of this system. For the initial system (157) $S_i = s_i = f_n$.

Thus, let us assume that a priori estimate of the "solutions" of system (158) is available in the form

$$X \in \mathfrak{X}_0 \subset R^m , \tag{159}$$

where $\mathfrak{X}_0$ is the given convex bounded set.

Let us assume that a priori estimate is given for parameter vector S of system (158)

$$S \in \mathfrak{S} \subset R^k . \tag{160}$$

Let us introduce the following

Definition 2. Let set $\tilde{\mathfrak{X}} \subset R^m$ of the solutions of equation (158) under condition (160) be taken to mean the set of such vectors X that vector S satisfying (160) can be found for each of them, such that (158) is met for the pair (X, S) .

The refined set $\mathfrak{X}_1$ of solutions of equation system (158) on condition that the first equation (158) as well as equations (159) and (160) are used will apparently take the form

$$\mathfrak{X}_1 = \tilde{\mathfrak{X}}_1 \cap \mathfrak{X}_0 . \tag{161}$$

Let us write the whole aggregate of solutions of equations (158) in compact form

$$\Phi(X,\bar{S}) = 0 , \tag{162}$$

where

$$\Phi^T(\cdot) = (\varphi_1(\cdot), \varphi_2(\cdot), \ldots, \varphi_N(\cdot)) , \quad \bar{S}^T = (S_1^T, S_2^T, \ldots, S_N^T) .$$

The refined set of solutions, as applied to system (162) considered as a sequence of equations (158), acquires the form of the recurrent sequence

$$\mathfrak{X}_i = \tilde{\mathfrak{X}}_i \cap \mathfrak{X}_{i-1} , \qquad i = 1, 2, \ldots . \tag{163}$$

The pay for the larger generality of the problem studied in the present section is that the estimates obtained hereinafter are generally only the "upper bound" estimates for the set of solutions, i.e. they comprise the exact set of the system solutions. In addition, the application domain of the obtained results narrows somewhat as compared to the solution obtained above in **Section 1.1.** , since the availability of a priori information $\mathfrak{X}_0$ is required in principle.

Set $\tilde{\mathfrak{X}}_1$ of solutions of (158) under condition (160) has the form of a "curvilinear strip" in the space of solutions and formally it can be represented in the form *

$$\tilde{\mathfrak{X}} = \{ X \mid \exists S \in \mathfrak{S} : \varphi(X,S) = 0 \} . \tag{164}$$

---

*Hereinafter, index i in equation (158) will be omitted to simplify the writing.

A large dimension of m , complex configuration of set $\mathfrak{X}_0$ and (or) the complexity of function $\varphi(\cdot)$ itself recommend decomposition for vector X . To this end, let us separate from the vector X set of its components Y with dimension p . Let us denote the remaining $r = m - pr = m - p$ components of vector X by vector Z . Thus, accurate to transformations $X^T = (Y^T, Z^T)$.

Let us assume that a priori information about the solution can be represented in the form $\mathfrak{X}_0 = \mathfrak{Y}_0 \times \mathfrak{Z}_0$ where $Y \in \mathfrak{Y}_0 \subset R^p$, $Z \in \mathfrak{Z}_0 \subset R^r$. Let us denote

$$\mathfrak{Y}_1 = \Pi_Y(\mathfrak{X}_1) \tag{165}$$

(respectively: $\mathfrak{Z}_1 = \Pi_Z(\mathfrak{X}_1)$) the projection of set $\mathfrak{X}_1 \subset R^m$ into subspace of "Y's" (respectively R^r is the subspace of "Z's"). Let us seek the upper bound estimate of set $\mathfrak{X}_1$ in the form of the Cartesian product of sets $\mathfrak{Y}_1 \times \mathfrak{Z}_1$ (ref. to **Fig. 11**). Determination of $\mathfrak{Y}_1$ by calculating (164) with the subsequent projection into subspace R^p represents in many cases a sufficiently complex problem. Because of this, let us take advantage of the projection method proposed below which carries out decomposition of the initial problem into a set of simpler problems.

Let us define set

$$\bar{\mathfrak{Y}} = \{ Y \mid \exists Z \in \mathfrak{Z}_0 , S \in \mathfrak{S} \mid \varphi(\cdot) = 0 \} . \tag{166}$$

Then the required estimate $\mathfrak{Y}_1$ is defined by the following

<u>Theorem 5.</u> (Method of projections). If a priori estimate of the solution $\mathfrak{X}_0$ of equation (158) under condition (160) has the form $\mathfrak{Y}_0 \times \mathfrak{Z}_0$, then the projection of solution $\mathfrak{X}_1$ into subspace R^p can be presented in the form

$$\mathfrak{Y}_1 = \bar{\mathfrak{Y}} \cap \mathfrak{Y}_0 , \tag{167}$$

where set $\bar{\mathfrak{Y}}$ is given by formula (166) .

Let us prove **Theorem 5** . <u>Necessity.</u> Let $\hat{Y} \in \mathfrak{Y}_1$. It follows from

(165) that $\exists\ \hat{Z} \mid (\hat{Y}, \hat{Z}) \in \mathfrak{X}_1$. It follows from (161) that

$$(Y, Z) \in \tilde{\mathfrak{X}} \tag{168}$$

and $(Y, Z) \in \mathfrak{Y}_0 \times \mathfrak{Z}_0$ which, in its turn, is equivalent to

$$\hat{Y} \in \mathfrak{Y}_0 , \tag{169}$$

$$\hat{Z} \in \mathfrak{Z}_0 . \tag{170}$$

It follows from (168) that $\exists\ \hat{S} \in \mathfrak{S} \mid \varphi\ [(\hat{Y}^T, \hat{Z}^T)^T, \hat{S}] = 0$. Since in this case (170) is true, (166) holds for $(\hat{Y}, \hat{Z}, \hat{S})$, i.e. $Y \in \bar{\mathfrak{Y}}$. From where jointly with (169) it follows that

$$\mathfrak{Y}_1 \subset \mathfrak{Y}_0 \cap \bar{\mathfrak{Y}} . \tag{171}$$

<u>Sufficiency.</u> Let now $\check{Y} \in \mathfrak{Y}_0 \cap \bar{\mathfrak{Y}}$. This means that

$$\check{Y} \in \mathfrak{Y}_0 , \tag{172}$$

$$\check{Y} \in \bar{\mathfrak{Y}} . \tag{173}$$

It follows from (172) and (166) that $\exists\ \check{Z} \in \mathfrak{Z}_0 ,\ \check{S} \in \mathfrak{S} \mid \varphi[(\hat{Y}^T, \hat{Z}^T)^T, \hat{S}] = 0$. Thus, vector $(\check{Y}^T, \check{Z}^T)^T$ with $\check{S}$ satisfies (164), i.e.

$$(\hat{Y}^T, \check{Z}^T)^T \in \mathfrak{X} . \tag{174}$$

The truth of (161), i.e. $(\check{Y}^T, \check{Z}^T)^T \in \mathfrak{X}_1$ follows from (174) and (172) and from the fact that $\check{Z} \in \mathfrak{Z}_0$. Therefore it follows by virtue of (165) that $\hat{Y} \in \bar{\mathfrak{Y}} \subset \mathfrak{Y}_1$. Thus,

$$\mathfrak{Y}_0 \cap \bar{\mathfrak{Y}} \subset \mathfrak{Y}_1 , \tag{175}$$

which jointly with (171) gives (167).

Thus, the method of projections gives the possibility to obtain the exact solutions of projection estimates by working in a smaller subspace.

The explicit description of set $\mathfrak{Y}_1$ in accordance with (167) involves some difficulties since the curvilinear strip (166) is restricted generally by nonlinear surfaces. Because of this, it is worthwhile to consider the estimates of set $\mathfrak{X}_1$, described by simpler surfaces, viz planes. To this end, let us proceed as follows. Equation (158) can be presented in ambiguous way in the form

$$a_1(X,S)y_1 + \dots + a_p(X,S)y_p - b(X,S) = 0 , \tag{176}$$

where y_i are components of vector Y, $i \in \overline{1,p}$, $a_i(X,S)$ and $b(X,S)$ - are, generally speaking, nonlinear functions of the both vector arguments.

Let us linearize this equation assuming its parameters to be interval ones. Let us define lower bound $\underline{a}_i$, $\underline{b}$ and upper bound $\overline{a}_i$, $\overline{b}$, estimates of the values of functions $a_i(\cdot)$ and $b(\cdot)$ over the whole domain of their definition ($\mathfrak{X}_0 \times \mathfrak{S}$) . Then the problem of finding the solution of equation (176) is reduced to a problem of the form (27), (28) where A and B are quantities specified on an interval. The solution of equations of this form was considered above in 1.1. It is generally a non-convex set $\hat{\mathfrak{Y}}$ in space R^p presented in the form of a union of a final number of convex polyhedra. Each of these polyhedra is described explicitly by a system of $p + 2$ linear inequalities which are specific for the polyhedron and p inequalities determine the respective orthant. Taking into account also the a priori estimate of the solution sought-for, i.e. set $\mathfrak{X}_0$, we obtain finally the upper bound estimate $\mathfrak{X}_1$ for the set of solutions $\mathfrak{X}_1$ of equation (158) with condition (160) in the following form (ref. to **Fig. 11**)

$$\hat{\mathfrak{X}}_1 = \hat{\mathfrak{Y}} \cap \mathfrak{X}_0 . \tag{177}$$

One remark should be made here. Since $\hat{\mathfrak{Y}}$ belongs to subspace R^p of space R^m, then intersection (177) is meant as the intersection of this subspace, i.e. without this remark, instead of (177), one should write

$$\hat{\mathfrak{X}}_1 = (\hat{\mathfrak{Y}} \times R^p) \cap \mathfrak{X}_0 .$$

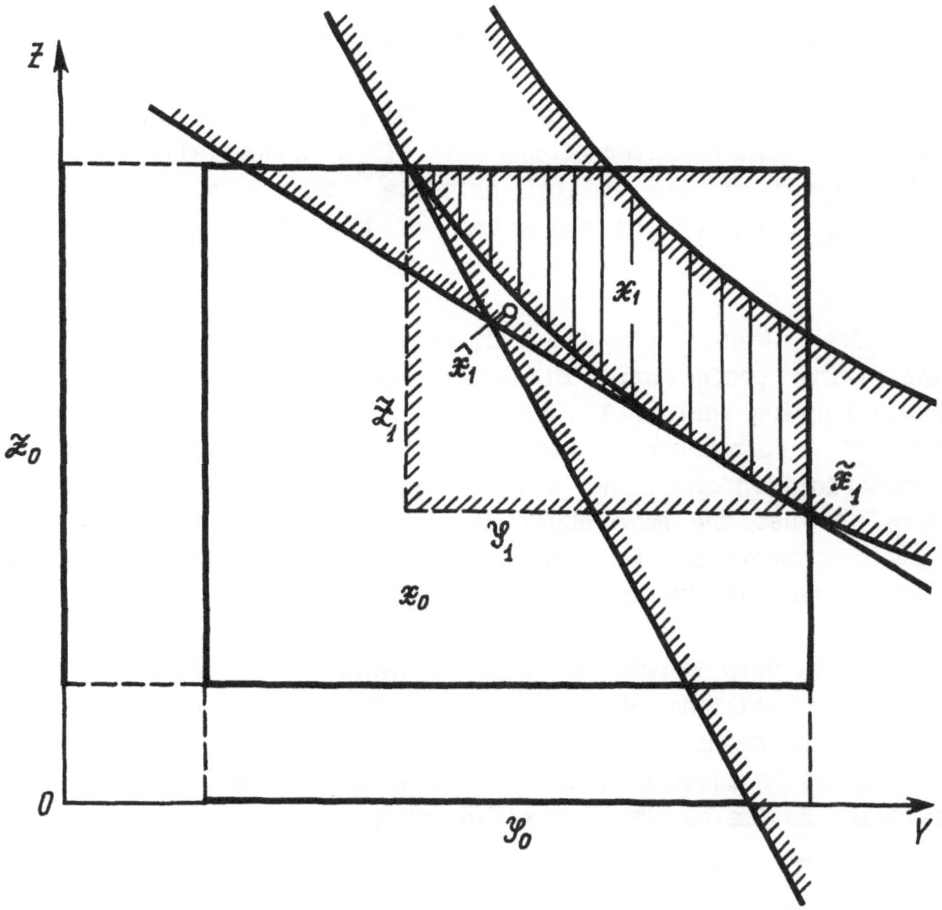

Fig. 11. Successive projection and linearization in constructing the set of solutions of nonlinear equation

From the practical point of view, it is convenient to realize the intersection procedure (177) in each orthant separately. In this form, the procedure itself is a union to one system of conditions describing $\mathfrak{X}_0$, with two inequalities defining the polyhedron and p inequalities defining the orthant being considered.

It should be noted also that the "dimensions" of the obtained estimate $\hat{\mathfrak{X}}_1$ (e.g. diameter of the set $s(\mathfrak{X}_1)$) essentially depend on the "dimensions" of the initial estimate $\mathfrak{X}_0$. In case of poor estimates $\mathfrak{X}_0$, the use of the method of linearizing projections is inefficient.

Let us draw attention to the arbitrariness existing in the method of linearizing projections in extracting from vector X the components forming vector Y . When this takes place, the refinement of only "Y" - components of vector X actually is carried out from the a posteriori information - from the constraint equation (158). It is natural to use the same equation also for the refinement of the remaining components by the method of linearizing projections . This is the main point of the method of <u>successive</u> linearizing projections.

Let us break down vector X into q sets of components Y^1, Y^2, ..., Y^q with dimensions p_1, ..., p_q , respectively, where $\Sigma_{pi} = m$. Let us apply the method of linearizing projections to equation (158) q times one after another, with extracting components Y^1, ..., Y^q , respectively. Estimates $\hat{\mathfrak{Y}}^i$, $i \in \overline{1,q}$ of the projections of the set of solutions $\mathfrak{X}_1$ onto subspaces R^{p_i} obtained in this case give, on the intersection with a priori information $\mathfrak{X}_0$ about the solution, the upper bound estimate

$$\hat{\mathfrak{X}}_1 = (\hat{\mathfrak{Y}}^1 \times \dots \times \hat{\mathfrak{Y}}^q) \cap \mathfrak{X}_0 \tag{178}$$

of the set of solutions $\mathfrak{X}_1$ in space R^m .

The main value of the method of successive linearizing projections described above consists, naturally, first of all in its linearizing potentialities. But another not less important aspect of this method is its "decomposing" property making it possible to overcome successfully the "dimensionality damnation". Let us illustrate this on the class of linear systems (27) on conditions (26). Let us note at the same time that the decomposing properties of the method of successive projections are well preserved with its application to nonlinear systems.

A rather efficient algorithm for finding exact set of solutions of a system of linear equations was already described above. Nevertheless, its practical application is restricted to comparatively low dimensions. Because of this, let us show the possibility to reduce the dimension of the problem by means of the method of successive linearizing projections what will be achieved at the cost of the upper bound coarsening of the set of solutions.

Strictly speaking, a solution of a system of N equations in m unknowns is representable in the form of a union of not more than 2^m polyhedra located in separate orthants and described by $(2N + m)$ inequalities with m arguments. The upper bound estimate of the set of solutions of the same system with the application of the method of successive linearizing projections is of the form of Cartesian product of sets $\mathfrak{Y}^1 \times \dots \times \mathfrak{Y}^q$. In this case, each set $\mathfrak{Y}^i$ is representable in the form of the union of not more than 2^{p_i} polyhedra each being located in separate orthant and described by $2N + p_i$ inequalities with p_i arguments. The following concrete example will illustrate the benefit from the application of this method.

Let $m=20$, $q=5$, $p_1=\dots=p_4=4$, $N=20$. In this case, the exact solution is described by a combination of a <u>million</u> (!) polyhedra each being set by 60 inequalities with 20 arguments which are very difficult to be handled from the computational point of view. But the upper bound estimate is described by 5 sets. When this takes place, each of the sets is formed only by 16 polyhedra each being described by 44 inequalities with 4 arguments.

Let us consider now the peculiarities of the formation of functions $a_i(\cdot)$ and $b(\cdot)$ for linear systems with already fixed set Y. Clearly there is a definite freedom of choice in constructing these functions since summand y_i (more precisely, the respective component x_k of vector X in the initial equation (158)) can be divided between $a_i(\cdot)y_i$ and $b(\cdot)$ with different weights. Let us find these most advisable weights.

Let us form type (176) equations from nonlinear equation (158) using two different methods and designating different functions with one and two primes, respectively. In this case, all functions coincide in the first and in the second method, except for $a_i(\cdot)$ and $b(\cdot)$. Grouping together common terms of these equations we designate them as $h(X,S)$. Then we obtain for the first and for the second method:

$$h(X,S) + a_i'(X,S)y_i - b'(X,S) = 0 , \tag{179}$$

$$h(X,S) + a_i''(X,S)y_i - b''(X,S) = 0 . \tag{180}$$

The difference between these equations consists in the fact that $a_i'(\cdot)$ is formed completely from the coefficient at y_i in the initial equation (158). Naturally, in this case $b'(X,S)$ does not contain terms with y_i. However, function $b''(X,S)$ in equation (180) contains summand with y_i with weight $(1 - \alpha)$ and the same summand with weight α $(0 < \alpha < 1)$ enters into function $a_i''(\cdot)$. Thus, the following relationship exists between coefficients of equations (179), (180)

$$a_i''(X,S) = \alpha a_i'(X,S) , \tag{181}$$

$$b''(X,S) = b'(X,S) + (1 - \alpha)a_i'(X,S)y_i . \tag{182}$$

Having applied the method of linearizing projections separately to equations (181) and (182), we obtain sets $\mathfrak{Y}'$ and $\mathfrak{Y}''$, respectively.

Lemma 3. Estimate $\mathfrak{Y}'$ is more exact than estimate $\mathfrak{Y}''$, i.e. $\mathfrak{Y}' \subset \mathfrak{Y}''$.

From this lemma directly follows

Theorem 6. Let linear equation (176) be given and set of components Y be fixed in the method of linearizing projections. Then the least for the inclusion estimates of the set of solutions will be given by such method of formation of functions $a_i(\cdot)$, $b(\cdot)$ in which $a_i(\cdot)$ are formed completely from coefficients x of the initial equation (158), which correspond to component y_i .

Let us prove **Lemma 3** . Let us consider the s-th orthant in the space of "Ys" . Without considering separately the space of "Zs" , let us find from (179) and (180) the upper and lower bounds of quantities $h(X,S)$, $a_i(X,S)$, $b(X,S)$ in the s-th orthant, marking the lower bounds by underlined bar and the upper bounds by overscribed bar. The method of determining the bounds for linearly entering and independent parameters is unambiguous and is carried out in accordance with usual laws of interval arithmetic (ref. e.g. to [21], [22]). Let us write down for each of two equations (179) and (180) the systems of inequalities in accordance with (35), (37) which determine in the s-th orthant the solutions of these equations $\mathfrak{Y}'$ and $\mathfrak{Y}''$, respectively

$$\mathfrak{Y}' : \left| \begin{array}{ll} \overline{h} + \overline{c}_i y_i \geq \underline{b}' , & (183) \\ \underline{h} + \underline{c}_i y_i \leq \overline{b}' , & (184) \\ [\, I - 2G(s)\,]Y \geq 0 , & (185) \end{array} \right.$$

$$\mathfrak{Y}'' : \left| \begin{array}{ll} \overline{h} + \alpha \overline{c}_i y_i \geq \underline{b}' - (1 - \alpha)\overline{c}_i y_i , & (186) \\ \underline{h} + \alpha \underline{c}_i y_i \leq \overline{b}' - (1 - \alpha)\underline{c}_i y_i , & (187) \\ [\, I - G(s)\,]Y \geq 0 , & (188) \end{array} \right.$$

where matrix $G(s)$ was defined above (ref. to (35)), and $\overline{c}_i = \overline{a}_i$ and $\underline{c}_i = \underline{a}_i$ when y_i is positive in the s-th orthant; $\overline{c}_i = \underline{a}_i$ and $\underline{c}_i = \overline{a}_i$ in the opposite case; $\overline{y}_i$ and $\underline{y}_i$ are the upper and the lower bounds for respective component x_k of the priori set $\mathfrak{X}$.

Let some vector be $\overset{*}{Y} \in \mathfrak{Y}'$. Suppose that the i-th component y_i satisfies (183). Then we have

$$\bar{h} + \alpha\bar{c}_i\overset{*}{y}_i \geq \underline{b}' - (1 - \alpha)\bar{c}_i\overset{*}{y}_i \geq \min_{y_i} \{ \underline{b}' - (1 - \alpha)\bar{c}_i y_i \} =$$

$$= \underline{b}' + \min_{y_i} \{ - (1 - \alpha)\bar{c}_i y_i \} = \underline{b}' - \max_{y_i} \{ (1 - \alpha)\bar{c}_i y_i \} =$$

$$= \underline{b}' - (1 - \alpha)\bar{c}_i y_i ,$$

i.e. $\overset{*}{y}_i$ satisfies (186). Having interchanged the positions of the overscribed and underlining bars as well as the minima and maxima and the inequality signs, we obtain in a similar way that it follows from the truth of (184) for $\overset{*}{y}_i$ the truth of (187). Thus, it follows from $\overset{*}{Y} \in \mathfrak{Y}'$ that $\overset{*}{Y} \in \mathfrak{Y}''$.

The proof of the theorem is carried out by a simple application of the lemma by induction.

Let us note that the amount of computations in analyzing the both equations is the same and does not depend on α .

Now, let us consider a special case of decomposition of linear systems with $q = m$ and $p = 1$, joining the method being considered with the interval estimation [29]. In this case, the obtained estimates $\mathfrak{Y}^i$, $i \subset \overline{1,m}$ will represent intervals. If the initial set $\mathfrak{X}_0$ is a polyhedron, then it is required at the initial step to solve m times the linear programming problem to find b^j ($j = \overline{0,1}$) . From this point on, determination of a_i^j and b^j is trivial and does not require computations. It follows from what has been said that the interval estimation provides the smallest computations, but only at the expense of the coarsening of the obtained estimates.

Thus, the method of successive projections functions on the class of linear equation systems functionsas a peculiar method of their decomposition, expanding and supplementing in this manner the well-known stores of decomposition means (e.g., ref. to [30]).

Let us apply the method of linearizing projections to nonlinear equation (158). The choice of the best dimension p and (with the chosen p) of the best in some sense set $\mathfrak{X}$ as well as the formation of the best functions Y and $a_i(\cdot)$, $b(\cdot)$ (with the chosen Y) is, generally speaking, an art. As to the choice of p , it is easy to show that the greater p is selected, the more accurate are the obtained estimates, however, the greater amount of computations is required to determine and to describe the estimates. The best accuracy is achieved with $p = m$. As regards the choice of Y and functions $a_i(X,S)$ and $b(X,S)$, it seems to be impossible to give here general recommendations. Nevertheless, we can consider individual subclasses of nonlinear equations for which some argued recommendations can be given. For example, the theorem for linear systems mentioned above can be generalized to a rather narrow class of nonlinear systems provided that restriction are imposed on the method of determining coefficients a_i^j , b^j .

Repeated applications of the method of projections with varying p, Y , $a_i(\cdot)$, $b(\cdot)$ (or some of them) enable the desired estimates to be refined successively. However, when the above quantities are not varied, then the repeated linearization gives no refinement, even though it may appear at first glance that this procedure can lead to success: in fact, the a priori sets do not change in this case.

Let us present the application of the above methods of projections on the class of bilinear equations. In order that the illustration of the application of the proposed procedures be most descriptive, let us resort to a simplest example from this class. Now, let a single equation be given

$$s_1x_1 + s_2x_2 + s_3x_1x_2 - s_4 = 0 \tag{189}$$

on condition that

$$\underline{s}_i \leqslant s_i \leqslant \bar{s}_i , \qquad i \in \overline{1,4} , \tag{190}$$

with a priori estimate of solution $\mathfrak{X}_0$ in the form

$$x_1 \in [\,\underline{x}_1, \bar{x}_1\,]\ , \quad x_2 \in [\,\underline{x}_2, \bar{x}_2\,]\ . \tag{191}$$

At first, let us assume that $p_1 = 1$ and set $y_1 = x_1$. Equation (176) taking the form in this case

$$a_1(X,S)y_1 - b(X,S) = 0\ , \qquad s^T = (\,s_1, s_2, s_3, s_4)\ , \tag{192}$$

will be generated in two methods, marked in Roman numerals

$$a_1^{I}(\cdot) = s_1\ , \qquad b^{I}(\cdot) = s_4 - (\,s_2 + s_3x_1\,)x_2\ , \tag{193}$$

$$a_1^{II}(\cdot) = s_1 + s_3x_2\ , \qquad b^{II}(\cdot) = s_4 - s_2x_2\ . \tag{194}$$

Using given estimates (191) under conditions (192), let us determine upper bounds $\bar{a}_1^{I}$, $\bar{b}^{I}$, $\bar{a}_1^{II}$, $\bar{b}^{II}$ and lower bounds $\underline{a}_1^{I}$, $\underline{b}^{I}$, $\underline{a}_1^{II}$, $\underline{b}^{II}$ of functions $a_1(\cdot)$ and $b(\cdot)$ of equation (192) in the both methods. Then we apply the method of linearizing projections in each method and obtain estimates $\hat{\mathfrak{Y}}^{I(1)}$ and $\hat{\mathfrak{Y}}^{II(2)}$, respectively. Their intersection makes it possible to obtain the estimate of the set of solutions for coordinate x_1

$$\hat{\mathfrak{X}}_1^{(1)} = \hat{\mathfrak{Y}}^{I(1)} \cap \hat{\mathfrak{Y}}^{II(1)} \cap [\,\underline{x}_1, \bar{x}_1\,]\ , \tag{195}$$

thus demonstrating the possibilities of repeated linearizing projections. Due to the symmetry of x_1 and x_2 in equation (189), having carried out the same operations over variable $y_2 = x_2$ and assuming $p_2 = 1$, we obtain estimates $\hat{\mathfrak{Y}}^{I(2)}$ and $\hat{\mathfrak{Y}}^{II(2)}$, respectively. These estimates present the method of successive linearizing projections enabling the remaining variable x_2 to be refined

$$\hat{\mathfrak{X}}_1^{(2)} = \hat{\mathfrak{Y}}^{I(2)} \cap \hat{\mathfrak{Y}}^{II(2)} \cap [\,\underline{x}_2, \bar{x}_2\,]\ . \tag{196}$$

Let us note that two methods of formation of type (192) equations do not exhaust all the potentialities of the repeated linearizing projections. Let us highlight the manifestation of both the linearizing properties of the method and of its decomposition property with $p_1 = p_2 = 1$.

Now, let us choose $p = 2$ to demonstrate the procedure of repeated linearization in a more complex but at the same time in more accurate version. In this case, the need in decomposition (191) of the initial conditions $\mathfrak{X}_0$ according to individual coordinates falls off, since the decomposition effect in absent. Let us assume that $y_1 = x_1$, $y_2 = x_2$. Then we obtain from (189) the equation

$$a_1(X,S)y_1 + a_2(X,S)y_2 - b(X,S) = 0 , \tag{197}$$

in which function $a_1(\cdot)$, $b(\cdot)$ is formed in four ways:

$$\text{(I)} \quad a_1(\cdot) = s_1 , \quad a_2(\cdot) = s_2 + s_3x_1 , \quad b(\cdot) = s_4 , \tag{198}$$

$$\text{(II)} \quad a_1(\cdot) = s_1 + s_3x_2 , \quad a_2(\cdot) = s_2 , \quad b(\cdot) = s_4 , \tag{199}$$

$$\text{(III)} \quad a_1(\cdot) = s_1 , \quad a_2(\cdot) = s_2 , \quad b(\cdot) = s_4 - s_3x_1x_2 , \tag{200}$$

$$\text{(IV)} \quad a_1(\cdot)= 0.5s_1, \quad a_2(\cdot)= s_2+ s_3x_1, \quad b(\cdot)= s_4- 0.5s_1x_1 . \tag{201}$$

Using the method of linearizing projections for each of the mentioned method, we obtain estimates $\hat{\mathfrak{Y}}^{I}$, ..., $\hat{\mathfrak{Y}}^{IV}$ whose intersection with a priori estimate $\mathfrak{X}_0$ gives the refined estimate $\hat{\mathfrak{X}}_1$, thus illustrating the application of the method of successive linearizing projections. Let us note that, due to the above lemma, the estimate $\hat{\mathfrak{Y}}^{IV}$ is non-informative since it comprises estimate $\hat{\mathfrak{Y}}^{I}$, i.e. the applications of the fourth method (once the first one has been already realized) is not worthwhile.

Digital simulation of the application of the methods of projections to bilinear equation (189) was carried out. In this case conditions (190) have been of the form: $24 \leqslant s_1 \leqslant 27$, $32 \leqslant s_2 \leqslant 36$, $96 \leqslant s_3 \leqslant 108$, $78 \leqslant s_4 \leqslant 88$, and a priori estimate (191): $x_1 \in [0; 1]$, $x_2 \in [0; 1]$. The simulation results are presented in **Fig.** 12 . The exact set of solutions of equation (189) $\mathfrak{X}$ is confined between two curves and it is shown jointly with a priori estimate in **Fig.** 12 by

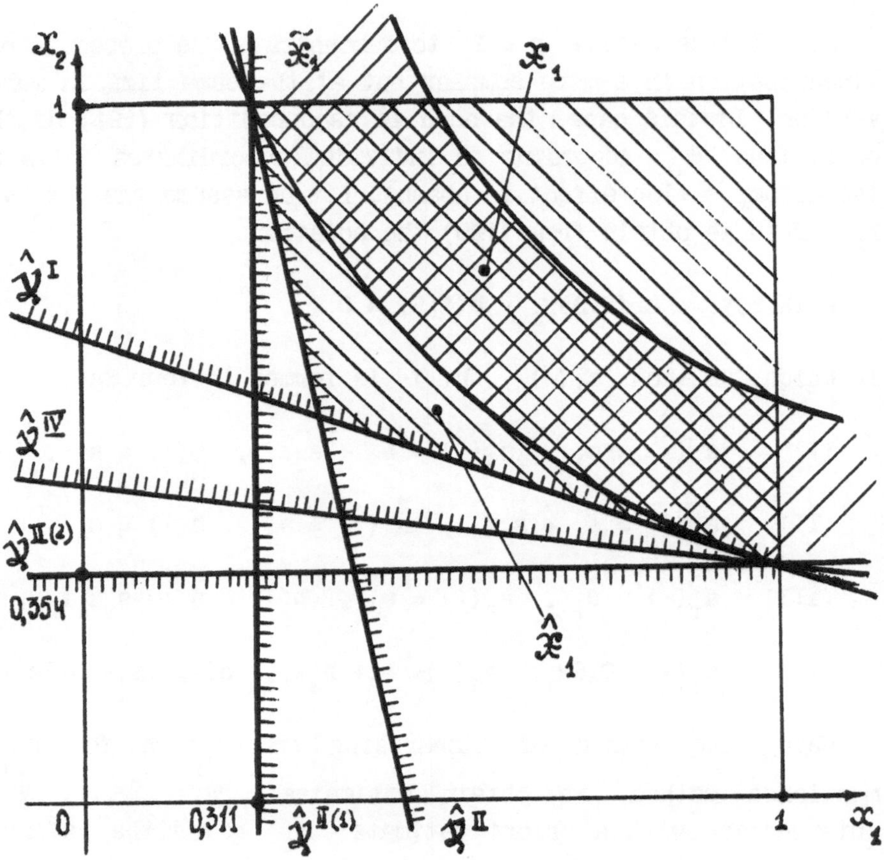

Fig. 12. Constructing the set of solutions of bilinear equation

cross-hatching sloping to the right. When $p_1 = 1$, $y_1 = x_1$ and set $\hat{\mathfrak{Y}}^{I(1)}$ defined by conditions (193) comprised a priori estimate $\mathfrak{X}_0$, and, therefore, it was non-informative which appears to be caused by the unsuccessful (for this estimate) choice of the initial conditions. A similar situation takes place for estimate $\hat{\mathfrak{Y}}^{I(2)}$ with $p_2 = 1$, $y_2 = x_2$. However, the estimates $\hat{\mathfrak{Y}}^{II(1)}$ from (194) and $\hat{\mathfrak{Y}}^{II(2)}$ turned out to be better, so that a posteriori estimates (195), (196) are optimal in the class of one-dimensional estimates and they take the form: $x_1 \in [\,0.311;\ 1\,]$, $x_2 \in [\,0.354;\ 1\,]$. Two-dimensional estimates $\hat{\mathfrak{Y}}^{I}$ and $\hat{\mathfrak{Y}}^{II}$ obtained with $p_2 = 2$ in accordance with (198), (199) enabled the estimate (195), (196) of the set of solutions of equation (189) obtained before to be refined. In this case, estimate $\hat{\mathfrak{Y}}^{III}$ turned out to be non-informative and $\hat{\mathfrak{Y}}^{IV}$, as one might expect, turned out to be worse than estimate $\hat{\mathfrak{Y}}^{I}$ (ref. to **Fig. 12**).

The final estimate $\hat{\mathfrak{X}}_1$ of the set of equations (189) is shown in **Fig.** 12 by cross-hatching sloping to the left.

The present example shows, in particular, that several repeated methods should be used to obtain as exact estimates as possible, i.e. there can be no optimal method among repeated methods.

In the foregoing, several methods have been described for obtaining upper bound estimates of the set of the possible solutions of linear and non-linear equations under conditions when parameters of these equations are set not by their true values but in the form of their membership in known sets. Some of these methods can be applied to one and the same initial problem and the obtained estimates can be both absorbed one by another (which means that the method with the more rough estimate was used in vain), and intersect (then the result of their intersection forms a new refined estimate). The both theorems presented here in some cases make it possible to estimate the expediency of the application of one or another method before its actual application. Thus, **Theorem 5** gives in a fixed subspace an optimal (unimprovable) upper bound estimate for the set of possible solutions of the equation (or system). In some cases, the space of solutions can be subdivided into such subspaces that the method of projections (166)-(167) giving optimal estimates can be easily realized in many of them. It is clear, that the use of the other repeated methods in these subspaces in unsuitable. **Theorem 6** makes it possible also to eliminate the useless realization of a number of methods, even though for a comparatively narrow class of problems.

CHAPTER 2

ANALYSIS AND SYNTHESIS OF ADAPTIVE CONTROL SYSTEMS

"The nature does not conform to the rules of the traditional probability".
R.Kalman

2.1.General Characteristic of Adaptive Control Systems

The parameter identification problem was considered above in **Chapter 1** as an independent problem being solved before constructing a control system for given plant. However, such a virtually heuristic method of decomposition of the problem of constructing a control system for a plant with the rather rough a priori estimates of its parameters into two independent problems (preliminary identification and synthesis of control system based on the obtained estimates of the controlled plant parameters) can be used by no means in all cases. The mentioned decomposition is impossible rather often in view of a number of reasons (in particular, because of the instability of the controlled plant) and it is necessary to combine in time both the process of control of the plant and the refinement of the estimates of its parameters in order to improve the quality of the control process when the original (a priori) estimates have been sufficiently coarse. It is common practice in the last decades to call this class of control systems adaptive systems. Many different approaches to their construction are known at the present time (e.g., ref. to [2], [31]-[39]). Nevertheless, one more approach to the construction of systems of this class will be described below whose distinguishing feature is among other things the use of guaranteed estimates of parameters with the procedure of their generation described above in sufficient detail.

Only the problem of optimal stabilization of a given plant will be mainly considered in the present chapter from a rather wide range of control problems existing at the present time. The reasons of such narrowing of the class of problems being considered are difficulties of the computing nature associated with the solution of the problem of optimal control synthesis under uncertainty conditions since the amount of calculations (which are to be carried out in real time) grows catastrophically fast with the growth of the set control interval for other classes of control problems, such as e.g. the problems of terminal control or synthesis of control systems optimal with respect to a functional.

Let us give a formal definition of the concept of "adaptive control system" itself whose definition at verbal level was introduced above. But first we should make several remarks of the preliminary nature.

Since two processes are realized simultaneously in adaptive control systems: the control itself and the identification, then it is obvious that some vector performance index should be introduced for objective evaluation of the quality of its functioning. One of its component reflect the quality of the course of the main control process and the quality of solution of the identification problem is evaluated by means of its second component.

Let controlled plant (77) be given with a priori estimate of its parameter vector in form (2), i.e. a class of such plants is given virtually, whose equation we shall write here once more for convenience of presentation:

$$X_{n+1} = A\ (L)X_n + B\ (L)u_n + Cf_n\ , \qquad n = 0,\ 1,\ 2,\ \dots\ , \tag{202}$$

where

$$A\ (\cdot) = \left|\begin{array}{c|c} 0 & I_{m-1} \\ \hline \multicolumn{2}{c}{\bar{L}^T} \end{array}\right|, \quad B\ (\cdot) = \left|\begin{array}{c} 0 \\ \dots \\ b \end{array}\right|, \quad C = \left|\begin{array}{c} 0 \\ \dots \\ 1 \end{array}\right|, \tag{203}$$

$$L^T = (\ \bar{L}^T,\ b) \in \mathfrak{L}_0\ , \tag{204}$$

$$f_n \in \mathfrak{f} = \{ f : | f | \leqslant \Delta = \text{const} \} . \quad (205)$$

Let the problem of optimal stabilization of this controlled plant be proposed for one-point set $\mathfrak{P}_0$ and empty set $\mathfrak{f}$, i.e. $\mathfrak{f} = \emptyset$. This problem is reduced to the necessity to minimize same given function $\omega_{n+1} = \omega\,(X_n, u_n, L)$ by selecting control u_n. Thus, for example, let Lyapunov function

$$v_n = v\,(X_n) = X_n^T P X_n \quad (206)$$

be introduced for controlled plant (202) at $\mathfrak{f} = \emptyset$ with quasi-diagonal matrix

$$P = \left\| \begin{array}{c|c} \underline{P} & 0 \\ \hline 0 & p_{mm} \end{array} \right\| , \quad p_{mm} > 0 \text{ is scalar function, } \underline{P}^T = \underline{P} > 0 \text{ is}$$

matrix with dimension $(m-1)\times(m-1)$. Then the first dimension of this function calculated along the trajectory of movement of system (202) is equal to

$$\Delta v_n = v_{n+1} - v_n = p_{mm} | \bar{L}^T X_n + b u_n |^2 + \bar{X}_n^T \underline{P} \bar{X}_n - X_n^T P X_n , \quad (207)$$

where

$$\bar{X}_n^T = (x_{2,n}, x_{3,n}, \ldots, x_{m,n}) .$$

Then the solution of the optimal damping problem, i.e. $\min \Delta v_n$ coincides with the solution of the problem of minimizing function

$$\omega_{n+1} = \omega\,(X_n, u_n, L) = p_{mm} | \bar{L}^T X_n + b u_n |^2 . \quad (208)$$

Another way of generating function $\omega_{n+1} = \omega\,(\cdot)$ can be shown on the example of construction of adaptive control system with reference model (ref. to, e.g., [33], [39]).

Let the objective of the control consist in minimizing at each n-th step the deviation of the m-th components of state vector X_n of system (202) and state vector $\tilde{X}_n$ of the reference system

$$\tilde{X}_{n+1} = \tilde{A}X_n , \tag{209}$$

where matrix $\tilde{A}$ has the same canonical structure as $A(\cdot)$ and vector $\bar{\tilde{L}}$ is equal to some preset (reference) value, i.e.

$$\bar{\tilde{L}} = \bar{\overset{*}{L}} , \tag{210}$$

Initial conditions for reference system $\tilde{X}_n$ at each n-th step are assumed to be equal to the initial conditions of system (202), i.e. $\tilde{X}_n = X_n$. Then

$$\omega_{n+1} = | \tilde{x}_{m,n+1} - x_{m,n+1} |^2 = | (\bar{\overset{*}{L}} - \bar{L})^T X_n + bu_n |^2 . \tag{211}$$

Let us note that if we set here $\bar{\overset{*}{L}} = 0$, which corresponds to the selection as reference system (209) of the system with the maximum speed of response, i.e. to the selection of $\lambda_i(\tilde{A}) = 0$, $i \in \overline{1,m}$ where $\lambda_i(\cdot)$ are eigen values of matrix $\tilde{A}$, then we obtain as a special case function ω_{n+1} determined by equation (208).

It is obvious that the number of different scalar functions ω_{n+1} by means of which the quality of the solution of the optimal stabilization problem is estimated can be also increased.

The quality of the solution of the problem of construction of the estimates of system parameter and state vectors was evaluated above by the diameters of these sets. Approximately equivalent from the "geometric" point of view but more labour consuming is the estimate by the volume $v_n^L = v(\mathfrak{L}_n)$ and (or) $v_n = v^X(\mathfrak{X}_n)$ of these sets.

Only diameters of these sets, i.e. $\delta_n^L = \delta(\mathfrak{L}_n)$ and $\delta_n^X = \delta(\mathfrak{X}_n)$ will be used in the present book for the estimation of sets $\mathfrak{L}_n$ and $\mathfrak{X}_n$. Then vector criterion

$$J_{n+1} = \begin{Vmatrix} \omega_{n+1} \\ \delta_{n+1}^L \\ \delta_{n+1}^X \end{Vmatrix} . \tag{212}$$

serves for the class of adaptive stabilization systems as the criterion for estimating the quality of their functioning.

It is obvious that also two-dimensional vector performance criteria can be used in individual cases for estimating the performance of adaptive stabilization system. Thus, for example, when state vector X_n is measured without noise, i.e. $Z = 0$ and, therefore, set $\mathfrak{X}_n$ is one-point one, then $\delta(\mathfrak{X}_n) = 0$. In this case the performance of the system can be evaluated by criterion

$$\underline{J}_{n+1} = \left\| \begin{matrix} \omega_{n+1} \\ \delta^{L}_{n+1} \end{matrix} \right\| . \tag{213}$$

On the other hand, if the a priori estimate of the parameter vector is sufficiently satisfactory and we can assume that $\delta(\mathfrak{C}_0) \approx 0$, then instead of (212) the value can be used

$$\bar{J}_{n+1} = \left\| \begin{matrix} \omega_{n+1} \\ \delta^{X}_{n+1} \end{matrix} \right\| . \tag{214}$$

Since the selection of control u_n at the n-th step influences in general case all three components of vector J_{n+1}, then, strictly speaking, the original (initial) control systems problem is reduced to the vector optimization problem for which constructive solution methods are absent at the present time. It is obvious that if sets $\mathfrak{C}_n$ and $\mathfrak{X}_n$ are not one-point ones, then the problem of minimizing index J_{n+1} is an ill-posed problem and it should be additionally defined in one or other way what will be done below, but now let us note that, in accordance with the existing at present practice of solving vector optimization problems (e.g., ref. to [40]-[45]) this problem is reduced in the long run to the scalar optimization problem, using one or other procedure of convolution of vector performance criterion J_{n+1} or changing over a part of components of vector J_{n+1} to the category of restrictions.

The introduced vector performance criterion of the control system operation makes it possible to introduce strict definitions of the concept of the control system "adaptability" following from the fact of its existence itself. This definition substantially differs from already known definitions having in the majority of cases only a descriptive character.

Definition 3. A system consisting of controlled plant (202), measuring device (119) and control device synthesizing control of class

$$u_n = u(\mathfrak{X}_n, \mathfrak{L}_n, \mathfrak{f}, \mathfrak{Z}) \tag{215}$$

from feasible area Ω (where $\mathfrak{X}_n$ and $\mathfrak{L}$ are the sequences of guaranteed a posteriori estimates of state and parameter vectors which are obtained in accordance with procedure described in detail in **Chapter 1**, $\mathfrak{f}$ and $\mathfrak{Z}$ are given non-empty sets - a priori estimates of disturbance f_n and measurement noise Z_n, respectively) is called weakly adaptive when there exists such instant of time $k < N$ that for all $n \geqslant k$ the inequalities are fulfilled

$$\omega_n < \omega_0 , \quad \delta_n^L < \delta_0^L . \tag{216}$$

Definition 4. We shall call adaptive such a weakly adaptive system for which at

$$\delta(\mathfrak{f}) = 0 \quad \text{and} \quad \delta(\mathfrak{Z}) = 0 \tag{217}$$

the condition are met:

$$\overline{\lim_{n\to\infty}}\ \omega_n = 0 , \quad \overline{\lim_{n\to\infty}}\ \delta_n^L = 0 . \tag{218}$$

Definition 5. We shall call robust such adaptive system for which at

$$\delta(\mathfrak{f}) \leqslant \Delta = \text{const} \quad \text{and} \quad \delta(\mathfrak{Z}) \leqslant \nabla = \text{const} \tag{219}$$

the conditions are met

$$\overline{\lim_{n\to\infty}}\ \omega_n \leqslant \bar{\omega}(\Delta, \nabla) = \text{const} , \tag{220}$$

$$\overline{\lim_{n\to\infty}}\ \delta_n^L \leqslant \bar{\delta}(\Delta, \nabla) = \text{const} \tag{221}$$

at any initial conditions and arbitrary realizations of disturbances f_n and Z_n from a given class (i.e., meeting conditions (204) and (205)).

Now, after a brief discussion of the main characteristics of the adaptive control systems, let us go over to a more detailed discussion of both the statement of the problem of synthesis of optimal (in the sense stipulated below) stabilization systems under uncertainty conditions and the methods of their solution.

2.2. Synthesis of Adaptive Optimal Stabilizing Systems (the simplest case)

Let us consider first the simplest problem of stabilizing system synthesis for the class of linear controlled plants (202) under condition that state vector X_n is measured exactly, i.e. that $Z = \emptyset$ and $f = \emptyset$.

Let us assume that a priori estimate (204) is given for parameter vector L , i.e. $L \in \wp_0$. In order to exclude from consideration the case $b = 0$ having no practical meaning, let us require in addition that set $\wp_0$ does not comprise point $b = 0$. For the sake of definiteness, let us assume hereinafter that $b > 0$ and, therefore, let us require that set $\wp_0$ does not comprise point $b \leqslant 0$.

Let the a priori estimate of parameter vector $\wp_0$ be rather coarse which excludes the possibility of obtaining satisfactory control results based on the use of only a priori estimate of parameters.

First we shall consider the case when no restrictions are imposed on control u_n . Let the initial control problem for one-point set $\wp_0$ consist in the minimization of given function $\omega_{n+1} = \omega(X_n, u_n, L)$

case $b = 0$ having no practical meaning, let us require in addition that set $\wp_0$ does not comprise point $b = 0$. For the sake of definiteness, let us assume hereinafter that $b > 0$ and, therefore, let us require that set $\wp_0$ does not comprise point $b \leqslant 0$.

Let the a priori estimate of parameter vector Ω_0 be rather coarse which excludes the possibility of obtaining satisfactory control results based on the use of only a priori estimate of parameters.

First we shall consider the case when no restrictions are imposed on control u_n. Let the initial control problem for one-point set Ω_0 consist in the minimization of given function $\omega_{n+1} = \omega(X_n, u_n, L)$, i.e. control u_n is sought for from the solution of the problem

$$\min_{u_n} \{ \omega(X_n, u_n, L) \} . \tag{222}$$

Since the value of vector L is given in this expression accurate only to its belonging to set Ω_0, it is obvious that this is an ill-posed problem and it should be redefined in one way or another. Let us invoke for this purpose the game approach to the statement of the problems of control under uncertainty conditions which is being intensively developed in the last few years (e.g., ref. to [46]-[50]) and let us assign to the nature (environment) generating all uncertain values the "insidious" intensions to maximize (within its stipulated potentialities) those performance indices of the control system which the system designer seeks to minimize. It is appropriate to note here that despite the criticism of such approach to the solution of control problems as extremely pessimistic one, as a matter of fact, there is no reasonable alternative in the situation being considered. Of course, except for giving up solving the problem at all.

Thus, let us look for control u_n at each n-th step out of the solution of the problem

$$\min_{u_n} \max_{L \in \Omega_n} \{ \omega(X_n, u_n, L) \} , \tag{223}$$

where

$$\omega_{n+1} = \omega(\cdot) = | \bar{L}^T X_n + b u_n |^2 . \tag{224}$$

In this case, as distinct from a non-adaptive system for which only a priori estimate Ω_0 would be used at each control step, we shall proceed here from the fact that the set-valued parameter identification procedure is used for generating the sequence of parameter estimates Ω_n considered in detail above, i.e. relationships **(84)** and **(85)** are used.

In the analysis of this set identification procedure, it was already found out that the linear independence of the sequence of vectors $\hat{X}_n^T = (X_n^T, u_n)$ is required for its efficient functioning. Deflecting our attention so far from the requirements to control sequence u_n dictated by the possibility to solve the problem of identification of controlled plant **(202)**, let us consider now the possibility to realize the passive identification mode when selecting controls u_n from the solution of problem **(223)**, **(224)**.

As is generally known, minimax problems come under the heading of the most difficult problems from the computing point of view. At present, a number of methods is already proposed for solving them (e.g., ref. to [45], [51]-[53]), but since problem **(223)**, **(224)** should be solved in real time, it must be admitted that universal methods of its solution known from literature prove to be inapplicable. A substantial gain in the amount of computations in determining the desired optimal control $\overset{*}{u}_n$ in problem **(223)**, **(224)** gives the following

<u>Theorem 7.</u> Optimal control $\overset{*}{u}_n$ providing the minimum of specific loss function **(224)** along the trajectories of the movement of system **(202)** at the optimal (for the "nature") value L_{opt} selected from set Ω_n is a root (moreover, the only root) of the equation

$$\varphi(u_n) = \max_{L\in\Omega_n} \{ \bar{L}^T X_n + bu_n \} + \min_{L\in\Omega} \{ \bar{L}^T X_n + bu_n \} = 0 . \qquad \textbf{(224)}$$

<u>Proof.</u> It will be recalled that the case is being considered when $\infty > b > 0$. Let us introduce functions

$$\varphi_1(u_n) = \max_{L\in\Omega_n} \{ \tilde{\varphi}(X_n, u_n, L) \} ,$$

$$\varphi_2(u_n) = \max_{L\in\Omega_n} \{ \tilde{\varphi}(X_n, u_n, L) \},$$

where

$$\tilde{\varphi}(\cdot) = \bar{L}^T X_n + bu_n .$$

It follows from the definition of function $\tilde{\varphi}(\cdot)$ that it is linear with respect to u_n, therefore functions $\varphi_1(\cdot)$ and $\varphi_2(\cdot)$ are continuous and strictly monotonically increasing functions. Indeed, function $\varphi_1(\cdot)$ is defined as the upper envelope of functions $\tilde{\varphi}(X_n, u_n, \overset{*}{L})$ linear in u_n, where $\overset{*}{L} \in \Omega_n$ takes on all its feasible values out of Ω_n. The "upper" envelope of these functions linear in u_n is a continuous and strictly monotonic function. The monotonicity and continuity of function $\varphi_2(\cdot)$ are proved in a similar manner. Since function $\varphi(\cdot)$ is the sum of continuous and strictly monotonically increasing functions $\varphi_1(\cdot)$ and $\varphi_2(\cdot)$, it possesses itself the same properties and, therefore, equation has one real root.

Let us show now that control $\overset{*}{u}_n$ being the root of equation (255) is the optimal control. We shall prove this by contradiction. For example, let instead of (225) the inequality take place at optimal control

$$\varphi(\overset{*}{u}_n) > 0 . \qquad (226)$$

It is obvious in this case that

$$\max_{L\in\Omega_n} \{ | \bar{L}^T X_n + bu_n | \} = \varphi_1(\overset{*}{u}_n) > | \varphi_2(\overset{*}{u}_n) | .$$

However, by virtue of the continuity and strict monotonicity of functions $\varphi_1(\cdot)$ and $\varphi_2(\cdot)$, there exists such

$$u_n = \overset{*}{u}_n + \Delta u_n ,$$

where Δu_n is a sufficiently small number at which

$$\varphi_1(u_n) < \varphi_1(\overset{*}{u}_n) \quad \text{and} \quad \varphi_1(u_n) \geqslant |\varphi_2(u_n)| .$$

From this it follows that

$$\underset{L \in \mathfrak{L}_n}{\max} \{|\bar{L}^T X_n + bu_n|\} = \varphi_1(u_n) < \varphi_1(\overset{*}{u}_n)$$

and, therefore, assumption (226) is false.

In a similar manner, we can show that at optimal control it is impossible to fulfill inequality $\varphi(\overset{*}{u}_n) > 0$ which is contradictory to (225). All this proves the validity of the theorem.

The property of function $\varphi(u_n)$ shown above makes it possible (despite the fact that it has a discontinuity of the first kind) to find the root of equation (225) by means of sufficiently simple iterative methods used to solve the problems of the similar kind (e.g., ref. to [54]-[56]).

It is quite appropriately to note here that the calculation of the value of function $\varphi(\cdot)$ and, therefore, of the values of functions $\varphi_1(\cdot)$ and $\varphi_2(\cdot)$ at any fixed value of u_n, required at one or other iterative method of solution of equation (225) is as a matter of fact the linear programming problem (only requirement $l_i \geqslant 0 \quad \forall i \in \overline{1,m}$ is absent in this case) and because of this it is to be solved by means of respective linear programming methods [40], [45], [57], [58]. Indeed, function ($\bar{L}^T X_n + bu_n$) takes at any fixed value of u_n its maximum (or minimum) value in one of the vertices L_n^i of polyhedron $\mathfrak{L}_n$ (ref. to **Fig. 13**).

When dimensionality of vector L is small then, in order to find the optimal control value, it proves to be advisable to reduce the problem of determining $\overset{*}{u}_n$ to the respective integer programming problem by introducing a final and also small number N of fixed levels $u_n^{(i)}$, $i \in \overline{1,N}$ which can assume control u_n and then to solve the latter problem, i.e. the problem

$$\underset{i \in \overline{1,N}}{\min} \quad \underset{j \in \overline{1,R}}{\max} \{|(\bar{L}_n^j)^T X_n + bu_n^{(i)}|\} ,$$

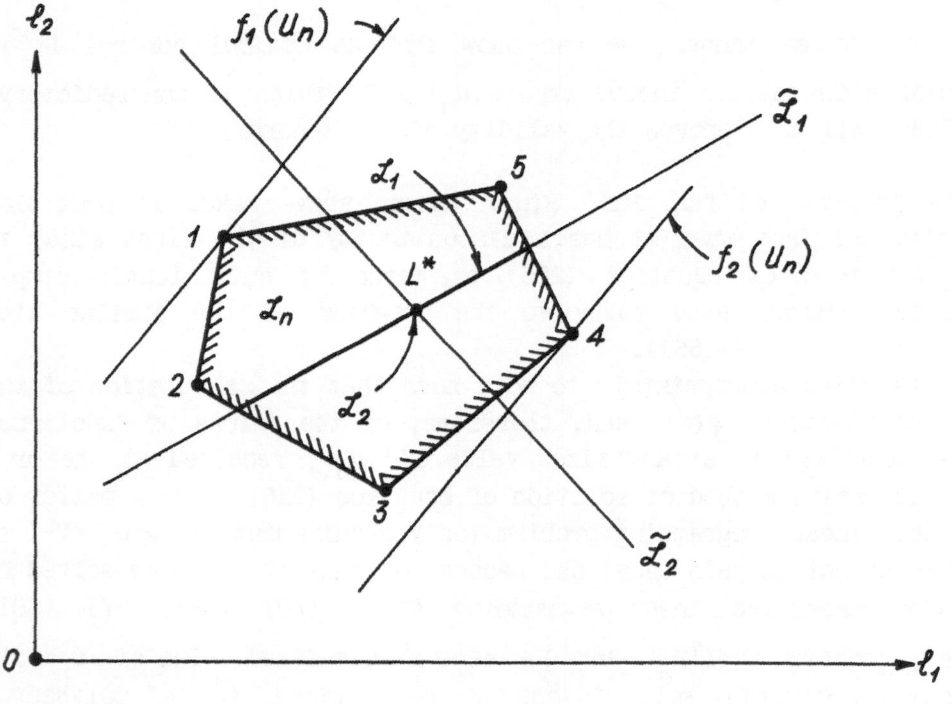

Fig. 13. Illustration of linear programming problem solution in determining optimal control

(where $L_n^j = [\ \overline{(L^j)^T},\ b\ ]$ is the j-th vertex of polyhedron $\wp_n$) either by means of respective algorithms for solving integer programming problems (e.g., ref. to [40], [58]) or by looking over all possible alternatives.

Let us consider now the same problem of control of the same plant **(202)** as above but with a restriction imposed on control u_n , i.e. under condition that

$$u_n \in \Omega = \{ u : \ | u | \leqslant c = const \} . \tag{227}$$

It is easy to show that in this case optimal control $\overset{o}{u}_n$ in problem **(223)**, **(224)** is defined as follows:

$$\overset{o}{u}_n = \left| \begin{array}{ll} \overset{*}{u}_n , & \text{if } \ | \overset{*}{u}_n | \leqslant c , \\ & \\ c \ \mathrm{sign}\ \overset{*}{u}_n , & \text{if } \ | \overset{*}{u}_n | > c , \end{array} \right\} \tag{228}$$

where $\overset{*}{u}_n$ is control defined by **Theorem 7** .

Let us analyze the functioning of the set identification procedure described above in **Chapter 1** as applied to plant **(202)** being considered here.

If each new measurement of form **(84)** defines a hyperplane whose position in space {L} differs from the positions of all preceding hyperplanes (conditions, under which this takes place will be defined below) then, since after each step the dimensionality of polyhedron $\wp_n$ decreases by one, the polyhedron $\wp_n$ degenerates after the expire of **(m + 1)** control steps to one-point set $\wp_n = \overset{*}{L}$ (ref. to **Fig. 13** in which the situation under consideration is presented for **m = 2**).

The solution of the problem **(223)**, **(224)** for one-point set (at **n ⩾ m +1**) (ref. to [59]) is optimal control

$$\overset{*}{u}_n = -\overset{*}{b}^{-1} (\overline{\overset{*}{L}})^T X_n , \quad n \geqslant m + 1 , \tag{229}$$

which provides the maximum speed of response of the synthesized control system.

Let us define now the conditions with whose observance the process of identification of the controlled plant parameters carried out simultaneously with the control process itself results in obtaining one-point set $\mathfrak{L}_n$. It was already noted above in **Chapter 1** that the use of the set intersection procedure is practically equivalent in the long run to the solution of linear equation system based on the sequential elimination of desired variables with the subsequent comparison of the obtained result with the a priori estimate $\mathfrak{L}_0$.

After (m + 1) cycles of control system operation functioning as described above, we obtain a system of observation equations (11). If matrix $\mathfrak{X}_{m+1}$ is nonsingular, i.e. if **det** $\mathfrak{X}_{m+1} \neq 0$, then we obtain from system (11) its solution in the form (12) and, therefore, the solution of the problem of system parameter identification terminates in obtaining the true value of vector **L** . The possibility to obtain this result is determined by the following

<u>Theorem 8.</u> a sequence of measurements

$$x_{m,n+1-i} = X_{n+i}^{T}\bar{L} + bu_{n+i} \ , \quad n = 0, 1, \ldots, \quad i = 0, 1, \ldots \ , \tag{230}$$

along the trajectory of motion of system (202) under control $\overset{*}{u}_n$ determined from the solution of problem (223) is linearly independent and, therefore, matrix $\mathfrak{X}_{m+1}$ of system (11) is non-singular.

For the proof of **Theorem 8** ref. to [10]. Matrix $\mathfrak{X}_{m+1}$ is non-singular when there in no eigenvalue equal to the zero among its eigenvalues $\lambda_i = \lambda(\mathfrak{X}_{m+1})$, $i \in \overline{1,m+1}$.

Let us prove the theorem by contradiction and let us assume that at least one of numbers λ_i is equal to zero. Then such non-zero vector S_i should exist which satisfies equation

$$\mathfrak{X}_{m+1}S_i = 0 \ . \tag{231}$$

Let us write S_i in the form

$$S_i^T = (\bar{S}_i^T, s_{i0}) . \tag{232}$$

Then we obtain from (231) and (232)

$$\left. \begin{aligned} X_n^T \bar{S}_i + s_{i0} u_n &= 0 , \\ X_{n+1}^T \bar{S}_i + s_{i0} u_{n+1} &= 0 . \end{aligned} \right\} \tag{233}$$

It follows from (233) that if $s_{i0} \neq 0$, then

$$u_{n+j} = -s_{i0}^{-1} X_{n+j}^T \bar{S}_i , \qquad j \in \overline{0,m} .$$

97

But it follows from (225) that

$$u_{n+j} \neq C^T X_n \quad \forall\ j \geqslant 0 , \tag{234}$$

which leads to contradiction.

Next let us consider the case $s_{i0} = 0$. If vector $\bar{S}_i$ is such that

$$\sigma_n = X_n^T \bar{S}_i = 0 , \tag{235}$$

then all other equations of the system (233) are not satisfied since it follows from (202) and (234) that

$$X_{n+1+j} \neq A X_{n+j} \qquad \forall\ j \geqslant 0 ,$$

where A is a numerical matrix of respective dimension. Therefore, vector $\bar{S}_i$ satisfying system of equations (233) does not exist also at $s_{i0} = 0$ which completes the proof of theorem.

Thus, in the mode of the optimal stabilization of the controlled plant (202) being considered now when control u_n is determined only from the condition of the minimization of specific loss function (234) this nevertheless does not prevent the simultaneous solution also of the other problem: the problem of controlled plant identification. In other words, in the case being considered, we succeed in combining the simultaneous solution of these two problems and as a result of this $\delta_n^L = \delta(\Omega_n) = 0$ and $\omega_n = 0$ after a final number of steps.

Thus, in accordance with **Definition 4** presented above, the system (202), (223) and (224) being considered is adaptive one.

In the case being considered, by virtue of the fact that with control $\overset{*}{u}_n$ determined from the solution of problem (223), (224) the both components of vector performance index $\underline{J}_{n+1}$ of the control system vanish in a final number of steps, there is clearly no need to reduce the minimax problem

$$\min_{u_n} \max_{L\in\Omega} \left\{ \underline{J}_{n+1} = \left\| \begin{matrix} \omega_{n+1} = \omega(X_n, u_n, L) \\ \delta_{n+1}^{L} = \delta[\Omega_{n+1}(u_n)] \end{matrix} \right\| \right\}$$

(for example, by means of linear convolution) to the scalar minimax control synthesis problem

$$\min_{u_n} \max_{L\in\Omega_n} \{ \sigma_{n+1} = \sigma(X_n, u_n, L, \alpha) \} .$$

Let us consider now an example illustrating the theorem formulated above and the statement stemming from it.

Let a special case of controlled plant (202) be given at $m = 1$ and $f = \emptyset$ ($\Delta = 0$), i.e. let

$$x_{n+1} = lx_n + bu_n , \quad x_0 = \overset{\circ}{x} , \quad n = 0, 1, 2, \dots . \tag{236}$$

Let us assume that a priori estimate Ω_0 of the vector $L^T = (l,b)$ is given by the system of inequalities

$$\underline{l} \leq l \leq \bar{l} , \qquad \underline{b} \leq b \leq \bar{b} , \tag{237}$$

where $\underline{l}$, $\overline{l}$ and $\underline{b}$, $\overline{b}$ are given numbers and $\underline{b} > 0$.

Let us assume that restriction (227) is imposed on control u_n , i.e. $| u_n | \leqslant c$.

In this case, specific loss function (224) has the form

$$\omega_n = x_{n+1}^2 = (lx_n + bu_n)^2 . \tag{238}$$

We shall solve optimal stabilization problem (223) at the following parameter values

$$\underline{l} = 1; \quad \overline{l} = 3; \quad \underline{b} = 1; \quad \overline{b} = 2; \quad c = 1; \quad \overset{\circ}{x} = 1 . \tag{239}$$

Let the true value of the parameter vector be equal to $\overset{*}{L}^T = (2, 1.5)$ (which corresponds to an unstable plant) and is not known to the system designer by the conditions of the problem.

According to **Theorem 7** , optimal control $\overset{*}{u}_0$ at the first step is determined from the solution of equation

$$f(u_0) = \max_{L \in \Omega_0} \{ lx_0 + bu_0 \} + \min_{L \in \Omega_0} \{ lx_0 + bu_0 \} = 0 . \tag{240}$$

Since it is obvious that $\overset{*}{u}_0 < 0$ at the parameter values set by inequalities (239), then

$$\max_{L \in \Omega_0} \{ l + bu_0 \} = \overline{l} + \underline{b}u_0 ,$$

$$\min_{L \in \Omega_0} \{ l + bu_0 \} = \underline{l} + \overline{b}u_0 .$$

Therefore equation (240) has the form

$$\overline{l} + \underline{b}u_0 + \underline{l} + \overline{b}u_0 = 0 , \tag{241}$$

from where we obtain that

$$\overset{*}{u}_0 = - \frac{\bar{l} + \underline{l}}{\underline{b} + \bar{b}} = - \frac{4}{3} \tag{242}$$

and since $| \overset{*}{u}_0 | > c$ then control $\overset{\circ}{u}_0 = c \text{ sign } \overset{*}{u}_0$ is realized, i.e. $\overset{\circ}{u}_0 = -1$.

Under the action of this control plant **(236)** changes from state $x_0 = 1$ to state $x_1 = 0.5$ and observation equation

$$l - b = 0.5 , \tag{243}$$

results from **(236)** which defines set $\tilde{\Omega}_1$ in the form of straight line.

The execution of the operation of intersection of sets $\tilde{\Omega}_1$ and Ω_0 gives a new estimate Ω_1 in the form of the straight line segment **(243)** restricted by two points L_1 and L_2 (ref. to **Fig. 14**) with the coordinates of vertices equal to $L_1^T = (1.5, 1)$, $L_2^T = (2.5, 2)$.

At $n = 1$, optimal control $\overset{*}{u}_1$ can be determined from the solution of equation

$$f(u_1) = \max_{L \in \Omega_1} \{ 0.5l + bu_1 \} + \min_{L \in \Omega_1} \{ 0.5l + bu_1 \} = 0 . \tag{244}$$

Since set Ω_1 is a straight line segment and there are only two extremum points then, therefore, if maximum is attained at $L = L_1$ then minimum take place at $L = L_2$ and vice versa. From this it follows that **(244)** can be rewritten in the form

$$0.75 + u_1 + 1.25 + 2u_1 = 0 .$$

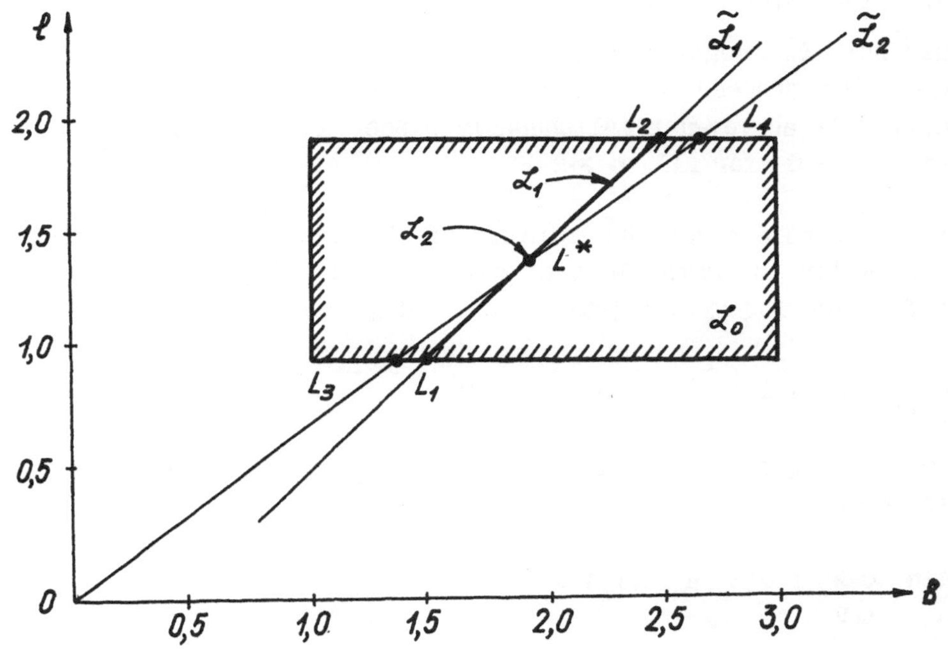

Fig. 14. Set parameter identification in the absence of noise

From where we obtain that

$$\overset{*}{u}_1 = 2/3 \ , \quad \text{i.e.} \quad |\ \overset{*}{u}_1\ | < c \ .$$

Under the action of control $\overset{*}{\tilde{u}}_1$, plant (236) changes over from state $x_1 = 0.5$ to state $x_2 = 0$. A new observation equation defines set $\overset{\sim}{\mathfrak{L}}_2$ in the form of straight line $0.51 - 2/3b = 0$.

From the intersection of sets $\overset{\sim}{\mathfrak{L}}_2$ and $\mathfrak{L}_1$ (ref. to **Fig. 14**) we obtain set $\mathfrak{L}_2$ containing only one point $\overset{*}{L}^T = (\ 2, 1.5)$. Thus, at $n = 2$ the process of identification is completed and the control objective is achieved simultaneously since $x_2 = 0$. At $u_n = 0$ we obtain $x_n = 0$ for all $n \geqslant 2$.

Let us consider now the solution of the problem of optimal control synthesis for the same class of with $\Delta \neq 0$. In this case specific loss function for system (202) instead (224) has the form

$$\omega_{n+1} = \omega(\cdot) = |\ \bar{L}^T X_n + b u_n + f_n\ |^2 \tag{245}$$

and we shall seek optimal control u_n respectively from the solution of the problem

$$\min_{u_n} \ \max_{\substack{L \in \mathfrak{L}_n \\ f_n \in \mathfrak{f}}} \ \{\ \omega(X_n, u_n, L, f_n)\ \} \tag{246}$$

with function $\omega(\cdot)$ defined by expression (245).

Let us show that in this case true is

<u>Statement 4.</u> Optimal control $\overset{*}{u}_n$ providing a minimum of specific loss function (245) on the trajectories of the motion of system (202) at $\mathfrak{f} \neq \emptyset$ is the unique solution of equation (225).

The truth of this statement follows from (226) and from the following relationship

$$\max_{\substack{L\in\Omega_n \\ f_n\in\mathcal{F}}} \{ |\, \bar{L}^T X_n + bu_n + f_n \,|^2 \} =$$

$$= \max_{L\in\Omega_n} \{ \Delta^2 + 2\Delta|\, \bar{L}^T X_n + bu_n \,|^2 + |\, \bar{L}^T X_n + bu_n \,|^2 \}. \tag{247}$$

It follows from (247) that the solution of problem (246), (246) is reduced to the solution of problem (223), (224).

Now, let us dwell at a greater length on the analysis of the properties of the system which controls plant (202), with control sequence $\overset{*}{u}_n$ determined from the solution of problems (223), (224) and set identification procedure (i.e., generation of estimates Ω_n) determined by relationships (84), (85). The existence of the non-empty set $\mathcal{F}$ results in qualitative changes of the synthesized system as compared to the case considered earlier where $\mathcal{F} = \emptyset$. In fact, since $f_n \neq 0$ then it is obvious that at any estimates Ω_n the asymptotic stability of system (202) with control $\overset{*}{u}_n$ determined from the solution of problem (223), (224) does not take place. Next, at $\mathcal{F} \neq \emptyset$ the sequence of estimates Ω_n generally does not degenerate to one-point set $\overset{*}{L}$ since the procedure of set identification (84), (85) generates some further unimprovable estimate Ω_n . Because of this, let us investigate asymptotic properties of controlled plant (202) with control determined from the solution of the equation of the form (225) with the use of the procedure of the refinement of the plant parameter estimates at each step described above.

It follows from (84) that equations defining the bounds of set $\tilde{\Omega}_{n+1}$ as the "hyperstrips" in parameter space have the form

$$\left.\begin{aligned} X_n^T\bar{L} + bu_n &= x_{m,n+1} - \Delta , \\ X_n^T\bar{L} + bu_n &= x_{m,n+1} + \Delta . \end{aligned}\right\} \tag{248}$$

It follows from (248) that if the estimate of parameters in the form of set Ω_n was used in generating control u_n then a measurement obtained after the realization of this control at the (n+1)-th step will be non-informative, i.e. set $\mathcal{P}_n$ will belong completely to set $\tilde{\mathcal{P}}_{n+1}$ and hence $\mathcal{P}_{n+1} = \mathcal{P}_n$ if and only if two inequalities are fulfilled simultaneously

$$\left.\begin{aligned} &\max_{L\in\mathcal{P}_n} \{ X_n^T\bar{L} + bu_n \} \leqslant x_{m,n+1} + \Delta , \\ &\min_{L\in\mathcal{P}_n} \{ X_n^T\bar{L} + bu_n \} \geqslant x_{m,n+1} - \Delta . \end{aligned}\right\} \qquad (249)$$

Having rewritten the last inequality in the form

$$- \min_{L\in\mathcal{P}_n} \{ X_n^T\bar{L} + bu_n \} \leqslant -x_{m,n+1} + \Delta$$

and using in this case the fact that control u_n satisfies equation (225), we obtain that the last of inequalities (249) takes the form

$$\max_{L\in\mathcal{P}_n} \{ X_n^T\bar{L} + bu_n \} \leqslant \Delta - x_{m,n+1} . \qquad (250)$$

And since it follows from (202) and (225) that

$$\max_{L\in\mathcal{P}_n} \{ X_n^T\bar{L} + bu_n \} \geqslant 0 ,$$

then we obtain finally that the last inequality in (249) can be fulfilled only when $x_{m,n+1} \leqslant \Delta$.

Similarly it can be shown that the first inequality in (249) can be fulfilled only when $x_{m,n+1} \geqslant -\Delta$.

Thus, to fulfill inequalities (249), it is necessary that inequality

$$| x_{m,n+1} | \leqslant \Delta \qquad (251)$$

is fulfilled, i.e. the observation equation obtained at some n will be informative only when the measured value $x_{m,n+1}$ obeys inequality (251). But if inequality (251) is not fulfilled then we obtain its improved estimate in accordance with the procedure of constructing the estimates of the parameter vector, i.e. $\Omega_{n+1} \subset \Omega_n$.

Let us dwell at greater length on the condition (251). Since

$$\max_{f_n \in f} \{ \mid X_n^T \bar{L} + bu_n + f_n \mid \} = \mid X_n^T \bar{L} + bu_n \mid + \Delta ,$$

then, therefore, if the nature generating disturbances f_n acts in the way most optimal for it, i.e. such value of f_n^* is realized at each step which maximizes the value of $\mid x_{m,n+1} \mid$, then inequalities (251) will not be fulfilled at each step when set Ω_n is not a one-point set containing only one point $L = \overset{*}{L}$. It follows from what has been said that in this case a refinement of the parameter vector estimate L will take place at each step despite the presence of additive disturbance f_n. Then on a lapse of a sufficiently large number of control steps at already known value L , we obtain from (223), (224)

$$u_n = - b^{-1} \bar{L}^T X_n \tag{252}$$

and, therefore, $x_{m,n+1} = f_n$. In this case, independently on the quality of the a priori estimate of parameter vector Ω_n , the synthesized control system is dissipative in the sense that beginning from some sufficiently large value $n = N$ inequality (251) is true. Function δ_n introduced above is taken as the criterion of the quality of the solution of the identification problem. However, if the nature generating disturbances f_n acts in the way non-optimal for it, i.e. it does not act in the way that we have postulated, then inequality (251) is fulfilled at all times beginning from some $n = N$ and as a result, the refinement of the parameter vector estimated available to that time can stop on some estimate which is then unimprovable. However, we obtain in this case from (251) and (249) that

$$\mid \bar{L}^T X_n + bu_n \mid \leqslant 2\Delta \qquad \forall\ n \geqslant N . \tag{253}$$

From this, a very important qualitative conclusion on the properties of the synthesized control system follows from this.

Statement 5. A system consisting of plant **(202)** with given estimates **(204),(205)** and control unit which synthesizes control U_n from the solution of equation **(225)** having regard to recursion relationship **(85)** where sets $\tilde{\Omega}_n$ represent "hyperstrips" in parameter space with bounds in the form of **(248)** determined as a result of measurements of the output coordinate $x_{m,n}$ of plant **(202)** with the use of its stored previous values forming vector X_{n-1} and of the value U_{n-1} is adaptive (in terms of **Definition 4**) and robust in terms of fulfillment of inequalities **(253),(251)** for sufficiently large **N**.

Let us noted that the adaptability of system **(202),(225)** follows from the fact that at $\Delta=0$ (i.e. at $f=0$) it coincides with system **(202),(223),(224)** for which (according to **Theorem 8**) the process of identification terminates in a final number of steps in obtaining a true value of parameter vector **L**.

In conclusion, let us dwell on a special case which is rather often encountered in control problems when it is known about disturbance f_n that it satisfies not only inequality **(205)** but also difference equation $f_{n+1} = f_n$, i.e. disturbance $f_n = \bar{f}$ is an unknown constant. As it is generally known (e.g., ref. to [60],[61]), to compensate the action of such kind of disturbances on the controlled plant, the technique of introduction of an integrator (summator) is widely used in control systems which imparts the property of zero-constant-error to the system. But this technique is heuristic one and it makes it possible only to obtain the required zero-constant-error properties of the control systems. That is why we shall not dwell on it and we shall consider the optimal solution of the synthesis problem in this case. Designating $\hat{L}^T = (\bar{L}^T, \bar{f})$ we obtain that the observation equations take the form

$$x_{m,n+1} = \hat{L}^T \hat{X}_n + bu_n + \mathring{f} , \tag{254}$$

where

$$\hat{X}_n^T = (X_n^T, 1) .$$

Thus, the problem of the refinement of the estimates of the system parameters is supplemented only with the problem of the refinement of the value of constant $\overset{\circ}{f}$ which can be treated as some additional parameter of system. The only feature of the problem of the refinement of the estimates of vector $\hat{L}$ values from (254) is that the order of the unknown parameter space increases by one.

It is easy to show that after m+2 steps the set identification procedure will generate one-point set Ω_n containing only one-point $\hat{L}^T = (\overset{*}{L}^T, \overset{\circ}{f})$. In this case the solution of the problem (233) has the form

$$\overset{*}{u}_n = - b^{-1} (L^T X_n + \overset{\circ}{f})$$

and this control provides the compensation of the action on the controlled plant of constant $\overset{\circ}{f}$ and besides it provides the asymptotic stability for the closed-loop control system as it was already shown earlier.

To illustrate the statements about the properties of adaptive control systems at $f_n = 0$ presented above, let us consider the simplest example of system (202) at $m = 1$, i.e. let us consider the system

$$x_{n+1} = l x_n + b u_n + f_n , \qquad n = 0, 1, 2, \dots . \tag{255}$$

A priori estimate Ω_0 is given for parameter vector $L^T = (l, b)$ defined by the following inequalities

$$\underline{l} \leqslant l \leqslant \bar{l} , \qquad 0 < \underline{b} \leqslant b \leqslant \bar{b} ,$$

where $\underline{l}$, $\bar{l}$ and $\underline{b}$, $\bar{b}$ are given numbers.

Let us assume the true value of parameter vector $\overset{*}{L}$ (unknown for the control system designer) to be equal to

$$\overset{*}{L}{}^T = (\bar{l}, \bar{b}) .$$

Let us assume then that set f $\forall n \geq 0$ is defined in the form

$$f = \{ f \mid | f | \leq 1 \} .$$

The objective of control is the minimization at each step of specific loss function

$$\omega_n = x^2_{n+1} = (lx_n + bu_n + f_n)^2 . \qquad (256)$$

On the basis of **Statement 5** optimal control $\overset{*}{u}_n$ is sought from the solution of equation

$$\varphi(u_n) = \max_{L \in \Omega_n} \{ lx_n + bu_n \} + \min_{L \in \Omega_n} \{ lx_n + bu_n \} .$$

The numerical simulation of the control process comprising a sequence of solutions of equations of form (246) with the simultaneous realization of the procedure of generating a sequence of guaranteed estimates Ω_n described above was carried out at parameter values $l = 1$, $\bar{l} = 3$, $b = 1$, $\bar{b} = 3$ and $x_0 = 1$ and the values of $f_n \in f$ have been as follows

$f_0 = 0.762$, $f_1 = -0.642$, $f_2 = 0.915$, $f_3 = 0.796$,

$f_4 = 0.697$, $f_5 = 0.620$, $f_6 = 0.564$, $f_7 = 0.526 \ldots$

The simulation results are presented in **Fig. 15** . A graph illustrating the variation of the quantity

$$\sigma_n = lx_n + bu_n$$

is also presented in this figure which characterized the accuracy of the control system operation, as well as the values of the root-mean-square error

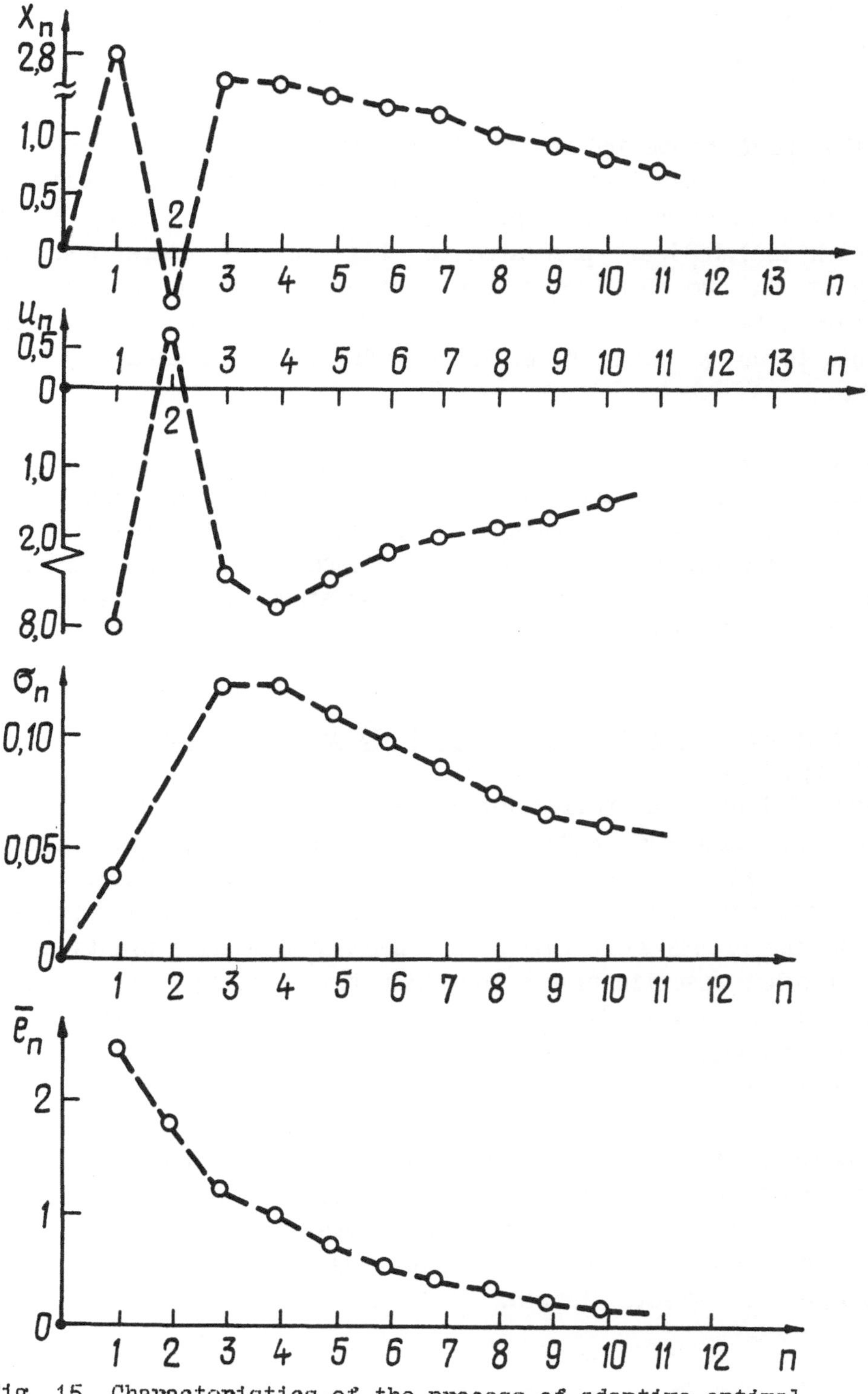

Fig. 15. Characteristics of the process of adaptive optimal with uncontrollable additive noise

$$\bar{l}_n = \frac{1}{n} \sum_{k=1}^{n} \sigma_k^2 .$$

Similar performance indices of the adaptive control system for the plant but obtained with the use of the recurrent objective inequalities (ROI) method proposed and developed by Prof. V. A. Yakubovich [62],[63] are presented in **Fig. 16** for comparison. The results of the numerical experiment simulating the operation of the ROI algorithm in controlling plant (255) have been kindly made available to us by Prof. V. A. Yakubovich and his colleague V. A. Bondarko.

The algorithms generated by the ROI method are simpler that algorithms proposed in the present book, however, their application (at least for the example being considered) gives the results (the quality of transient processes in a closed-loop adaptive control system) which are worse than the results obtained with the use of the control algorithms described above.

2.3. Synthesis of Adaptive Optimal Stabilizing Systems (the General Case)

A class of linear controlled plants was considered above which are described by difference equation (202) in which among other things matrix **A**(·) has canonical form (a form of Frobenius matrix). This class of discrete dynamic systems does not comprise such important plants as the plants with time delay, non-minimum-phase plants and a number of other specific plants for which the solution of the problem of their stabilizing control involves a variety of difficulties even if their parameters are completely known. The actuality of these control problems was repeatedly pointed out.

It should be noted that generally speaking, difference equations of the form (202) are of course very convenient for the analysis and synthesis but they can not be always obtained directly from the original, scalar difference equation. Because of this, let us dwell on this problem at a greater length.

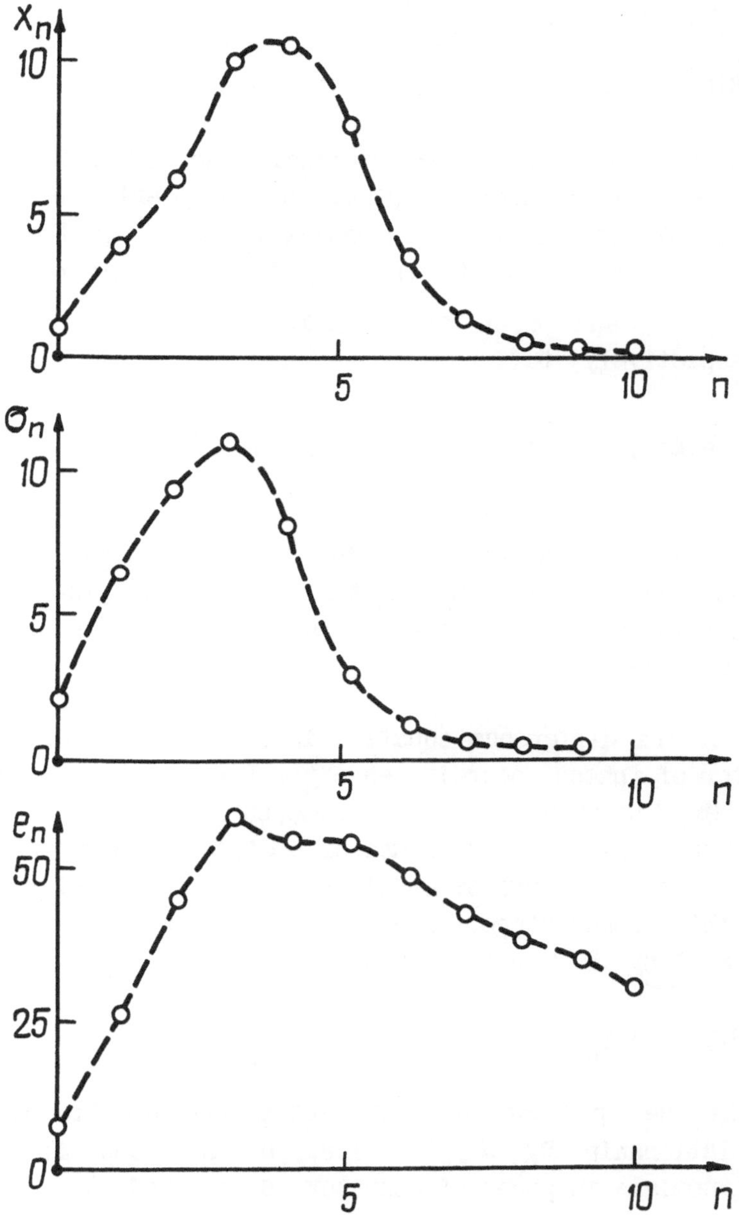

Fig. 16. Characteristics of the process of adaptive control (ly Yakubovich) with additive noise

Let the scalar difference equation of the form

$$\bar{A}(E)y_n = \bar{B}(E)u_n + f_{m+n} , \quad n = 0, 1, 2, \ldots , \tag{257}$$

play the role of the original mathematical model of the controlled object where y_n is the scalar "output" of the plant, u_n and f_n are scalar control and disturbance, respectively, E is the operator of the shift (displacement) "to the right" defined by relationship $Ey_n = y_{n+1}$, $\bar{A}(\cdot)$ and $\bar{B}(\cdot)$ are polynomials in power of E which are equal, respectively, to

$$\bar{A}(\cdot) = \sum_{i=0}^{m} \bar{a}_i E^i , \qquad \bar{B}(\cdot) = \sum_{j=0}^{s} \bar{b}_j E^j , \quad s < m .$$

Recall that this is the form in which difference equations of the motion of the plants continuous by their nature are obtained when using the apparatus of the discrete Laplace transform (z -transform) (e.g., ref. to [64]-[67]).

This form of the difference equation is inconvenient in the sense that the vector of phase coordinates $Y_n^T = (y_n, y_{n+1}, \ldots, y_{n+m-1})$ will contain in this case $(m - 1)$ components inaccessible for the measurement. Therefore let us make use of a different structure (form) of the phase space of the system considered above in **Chapter 1** which is free from this disadvantage. To do this, let us multiply both sides of equation (257) by E^{-m} which gives

$$A(D)y_n = B(D)u_n + f_n , \tag{258}$$

where D is the operator of shift (displacement) "to the left" defined by relationship $Dy_n = y_{n-1}$, i.e. $D = E^{-1}$ and $A(\cdot)$ and $B(\cdot)$ are polynomials in power of operator D defined as

$$A(D) = \sum_{i=0}^{m} a_i D^i , \qquad B(D) = \sum_{j=0}^{s} b_j D^j .$$

Equation (258) can describe in the general case both unstable plants (if among the roots of equation $A(D) = 0$ there are roots D_i less then 1 in absolute value) and non-minimal-phase plants i.e. the plants for which a parts of the roots of equation

$$B(D) = 0$$

lies within the unit circle. This circumstance prevents the problem of control synthesis from being reduced to the problem of control synthesis for the plant

$$A(D)y_n = \tilde{u}_n + f_n$$

by the change of variables

$$B(D)u_n = \tilde{u}_n \tag{259}$$

and then from determining the desired value u_n already in terms of the found value $\tilde{u}_n$. The impossibility to do this follows from the fact that the general solution of inhomogeneous equation **(259)** will comprise a divergent solution of the respective homogeneous equation.

Clearly the lack of satisfactory estimates of the parameter vector e.g. for a non-minimum-phase controlled plant by no means alleviates the problem of their control. Because of this it is advisable to dwell at least briefly on some aspects of control synthesis for the controlled plants of the general form described by vector-matrix equations

$$\tilde{X}_{n+1} = \tilde{A}X_n + \tilde{B}u_n + \tilde{D}f_n , \quad n = 0, 1, 2, \dots , \tag{260}$$

assuming so far that all elements of matrix $\tilde{A}$ and vectors $\tilde{B}$ and $\tilde{D}$ are unknown to the system designer. Recall that in accordance with the results presented in **Appendix A** , matrix $\tilde{A}$ and vector $\tilde{B}$ in this equation already have no canonical form (ref. to expressions **(A.8)**).

Let us place one rather natural requirement on the parameters of plant **(260)**: the requirement of its controllability. Then, as it is shown in **Appendix A** , there exists such linear transformation **(A.11)** bringing systems of equation **(260)** to the form **(A.13)** in which matrix $\check{A}$ and vector $\check{B}$ already have canonical structure similar to structure **(78)**. Therefore, there exists control u_n making the right-hand side of the last of equations **(A.13)** vanish thus providing the maximum speed of response of the closed-loop control system (in the absence of disturbance f_n).

Recall that as it was mentioned in **Appendix A** , this conclusion on the solution of the problem of the maximum speed of response is true in particular also for non-minimum-phase controlled plants. Thus, in solving the optimal stabilizing problem for controlled plant (260), specific loss function (ref. to (A.13)) has the form

$$\omega_n = | C^T\tilde{A}^{m+s-1}\tilde{X}_n + C^T\tilde{A}^{m+s-2}\tilde{B}u_n + C^T\tilde{A}^{m+s-2}\tilde{D}f_n | , \tag{261}$$

where $C^T = (c_1, c_2, \ldots, c_{m-1}, 1, c_{m+1}, \ldots, c_{m+s-1})$.

Components c_j ($j \neq m$) of vector C are determined from the solution of the linear equation system (A.10), i.e.

$$C^T\tilde{A}^{j-1}\tilde{B} = 0 , \qquad j \in \overline{1,m+s-2} . \tag{262}$$

Then it is easy to show that at $f_n \equiv 0$ optimal control $\overset{*}{u}_n$, i.e. control which provides the minimum of specific loss function ω_n has the form

$$\overset{*}{u}_n = - R^T\tilde{X}_n , \tag{263}$$

where

$$R^T = (C^T\tilde{A}^{m+s-2}\tilde{B})^{-1}C^T\tilde{A}^{m+s-1} .$$

It will be recalled here that the controllability condition

$$C^T\tilde{A}^{m+s-2}\tilde{B} \neq 0 \tag{264}$$

for plant (260) is assumed to be fulfilled.

Let us consider now the solution of the same optimal stabilizing problem for the class of plants (260) assuming that only a priori estimate of the form (204), i.e.

$$\tilde{L} \in \tilde{\mathfrak{L}}_0 , \tag{265}$$

where $\tilde{\Omega}_0$ is given convex polyhedron in space $\{\tilde{L}\}$, is known about the parameter vector of the plant

$$\tilde{L}^T = (a_1, a_2, \ldots, a_m, b_1, b_2, \ldots, b_s) . \tag{266}$$

Let us assume in addition to condition (266) that also for the true value of parameter vector $\tilde{L}$ the condition

$$\tilde{\gamma}(\tilde{L}) = C^T\tilde{A}^{m+s-2}\tilde{B} \neq 0$$

is fulfilled, i.e. that the plant is controllable. Let us assume without the loss of generality that

$$\tilde{\gamma}(\tilde{L}) = C^T\tilde{A}^{m+s-2}\tilde{B} > 0 . \tag{267}$$

As related to disturbance f_n, let us retain valid the above assumption that a priori estimate (3) is known for it, i.e.

$$f_n \in \mathcal{f} \qquad \forall\, n \geqslant 0 , \tag{268}$$

where set $\mathcal{f}$ is defined by inequality $| f | \leqslant \Delta$. Then, without repeating again all what was said about the refinement of the values of $\tilde{L}$ and f_n, let us formulate the problem of synthesis of the optimal (in terms of the minimum of specific loss function ω_n) control in the form of the following minimax problem

$$\min_{u_n \in \Omega} \ \max_{\tilde{L} \in \Omega_n} \ \max_{f_n \in \mathcal{f}} \ \{ \omega_n = \omega(\tilde{X}_n, u_n, \tilde{L}, f_n) \} , \tag{269}$$

where ω_n is a function defined by equations (261); Ω is the set of feasible control; $\tilde{\Omega}_n$ is a sequence of estimates of parameter vector $\tilde{L}$ with the procedure of their construction being described above in **Chapter 1** in sufficient detail.

We shall consider hereinafter mainly the problem of the optimal control synthesis without restrictions on control.

Let us introduce designation

$$\xi(\tilde{X}_n,\tilde{L},f_n) = C^T\tilde{A}^{m+s-1}\tilde{X}_n + C^T\tilde{A}^{m+s-2}\tilde{D}f_n \ . \tag{270}$$

Then (with regard to designation (267) introduced earlier) let us present specific loss function (261) in the form

$$\omega_n = |\ \xi(\tilde{X}_n,\tilde{L},f_n) + \tilde{\gamma}(\tilde{L})u_n\ | \ . \tag{271}$$

Since the expression under the sign of modulus is linear in u_n and in accordance with what has been said above $\tilde{\gamma}(\cdot) > 0$, then in optimal control $\overset{*}{u}_n$ determined from the solution of problem (269) true is the following equality*)

$$\max_{\tilde{L}\in\tilde{\Omega}_n} \max_{f_n\in f} \{ \xi(\tilde{X}_n,\tilde{L},f_n) + \tilde{\gamma}(\tilde{L})\overset{*}{u}_n \} = \\ = - \min_{\tilde{L}\in\tilde{\Omega}_n} \min_{f_n\in f} \{ \xi(\tilde{X}_n,\tilde{L},f_n) + \tilde{\gamma}(\tilde{L})\overset{*}{u}_n \} \ . \tag{272}$$

It is easy to show that the following theorem is valid which is an analog of **Theorem 7**

<u>Theorem 9.</u> Let controlled plant (260) be given for which condition (264) is fulfilled, with matrix $\tilde{A}$ and vectors $\tilde{B}$ and $\tilde{D}$ defined by expressions (A.8). Estimate of the form (265) is known about parameter vector $\tilde{L}$ (given by expression (266)) and estimate (268) is true for disturbance f_n . Then optimal control $\overset{*}{u}_n$ at each step for specific loss function (271) is the root (the unique one) of the equation

$$\check{\varphi}(u_n) = 0 \ , \tag{273}$$

---

*) The reasonings substantiating the truth of this equality are virtually the same as in the proof of the validity of **Theorem 7** and because of this they are not presented here.

where

$$\check{\varphi}(u_n) = \max_{\tilde{L}\in\tilde{\Omega}_n} \{ C^T\tilde{A}^{m+s-1}\tilde{X}_n + \tilde{\gamma}(\tilde{L})u_n + \delta| C^T\tilde{A}^{m+s-2}\tilde{D} | \} +$$

$$+ \min_{\tilde{L}\in\tilde{\Omega}_n} \{ C^T\tilde{A}^{m+s-1}\tilde{X}_n + \tilde{\gamma}(\tilde{L})u_n - \delta| C^T\tilde{A}^{m+s-2}\tilde{D} | \} , \qquad (274)$$

and $\tilde{\Omega}_n$ is a sequence of estimates determined by expressions of the form (85).

Let us dwell now on some features of the procedure of identification of parameter vector $\tilde{L}$ for plant (260). According to expressions (A.8) for plant (260), the observation equations which define set $\tilde{\Omega}_n$ have the form

$$x_{n+1} = - \sum_{i=1}^{m} a_i x_{m-i+1,n} + \sum_{i=1}^{s} b_i x_{m+s-i+1,n} + b_1 u_n + f_n , \qquad (275)$$

where $x_{j,n}$, $j \in \overline{1,m+s}$ are components of state vector $\tilde{X}_n$.

Since equations (260) as well as control (77) are linear with respect to parameter vector, then, because (260) and (77) coincide within designations, all what has been said above about the methods of realization of the recursion procedure defining the sequence of estimates $\tilde{\Omega}_n$ and about the properties of this sequence is fully applicable also to the case being considered here. In view of this, the synthesized control system is adaptive in terms of **Definition 4** if the quality of identification is evaluated by the diameter of the set $\delta_n = \delta(\tilde{\Omega}_n)$.

But if disturbances f_n satisfy additional condition (52) and this circumstance is taken into account in realizing the procedure of the parameter vector identification then the control system for the plant (260) synthesized here will be robust. Since the proof of the validity of this statement is carried out in accordance with procedure completely similar to that used in **Section 1.2.** , there is no need for its description here.

Let us emphasize one important circumstance which is of principal importance. Since no other restrictions are imposed on the relationship between parameters a_1 and b_1 of the parameter vector of controlled plant $\tilde{L}$ except for the quite natural requirement of its controllability, then it is clear that the conclusions made here about the synthesis of adaptive and strongly adaptive control systems for the class of plants **(260)** hold true also in the special case when the plant is non-minimum-phase one.

Thus, the problem of synthesis of adaptive optimal stabilizing systems for linear non-minimal-phase controlled plants should be taken to be solved (within the framework of the problem statement presented here).

Now, let us dwell on the analysis of some qualitative properties of the closed-loop control systems synthesized in accordance with the technique presented here. To do this, let us consider some special cases of the control problem being analyzed. If set $\mathfrak{f}$ is empty, i.e. disturbances f_n are absent, then it follows from all what has been said here and in **Section 1.2.** that parameter vector $\tilde{L}$ is identified completely in a final number of steps, i.e. for $n = m + s$ set $\tilde{\Omega}_n$ changes to one-point one containing only one point $\tilde{L}^*$. Then from the solution of problem **(269)** we obtain control **(263)** which brings asymptotic stability to the closed-loop system and what's more which provides the process of a final duration.

Now let set $\mathfrak{f}$ be non-empty and, besides, disturbances f_n satisfy additional condition **(52)**. Then the application of the identification procedure being considered here (with the use of the algorithm for the solution of this problem described in **Section 1.1.** and oriented to the application of computer) provides the determination of the parameter vector of the controlled plant up to arbitrary accuracy, i.e. we can assume that at $n \geqslant N$ set $\tilde{\Omega}_n$ will be a single-point set containing only one point $\tilde{L} = \tilde{L}^*$. Then optimal control u_n^* at $n > N$ is defined by the expression **(263)** and as it follows from **(A.13)**

$$z_{m+s-1,n} = C^T\tilde{A}^{m+s-2}\tilde{D}f_{n-1} \qquad \text{at } n > N . \tag{276}$$

Using expression **(A.13)**, let us determine now the values of the remaining components $z_{i,n}$ of vector Z_n after m following steps at $n > m + N$

$$z_{i,n} = \sum_{k=1}^{m+s-i} C^T\tilde{A}^{k+i-2}\tilde{D}f_{n-k} , \quad i \in \overline{1,m+s-1} . \tag{277}$$

It follows from **(276)** and **(277)** that components of vector Z_n are restricted in asymptotics and the value of this restriction is determined by the value of the restriction imposed on f_n , i.e. by the value of Δ (ref. to expression **(268)**), as well as by the true parameter values of plant **(260)**.

Since vectors Z and X_n are related by linear non-degenerate transformation **(A.11)** which can be rewritten in the form

$$\tilde{X}_n = H^{-1}Z_n , \tag{278}$$

where matrix H is given by expression **(A.12)**, then the boundedness of components of vector $\tilde{X}_n$ follows from the boundedness of the components of vector Z .

Thus, the control system synthesized in accordance with **Theorem 9** under conditions stipulated above is robust in terms of **Definition 4** and, therefore, it is also dissipative. To obtain the dissipativity estimates of the system, let us substitute expressions **(277)** into **(278)** and we obtain

$$\tilde{X}_n = H^{-1}\tilde{H}\tilde{F}_n , \tag{279}$$

where

$$\tilde{F}_n = (f_{n-(m+s-1)}, f_{n-(m+s-2)}, \ldots, f_{n-2}, f_{n-1}) ,$$

$$\tilde{H} = \left\| \begin{array}{c|c|c|c|c} C^T\tilde{A}^{m+s-2}\tilde{D} & C^T\tilde{A}^{m+s-3}\tilde{D} & \dots & C^T\tilde{A}\tilde{D} & C^T\tilde{D} \\ \hline 0 & C^T\tilde{A}^{m+s-2}\tilde{D} & \dots & C^T\tilde{A}^2\tilde{D} & C^T\tilde{A}\tilde{D} \\ \hline \vdots & \vdots & \dots & \vdots & \vdots \\ \hline 0 & 0 & \dots & \dots & C^T\tilde{A}^{m+s-3}\tilde{D} \\ \hline 0 & 0 & \dots & 0 & C^T\tilde{A}^{m+s-2}\tilde{D} \end{array} \right\| . \qquad (280)$$

It follows from these expressions that the estimate of dissipativity of an arbitrary coordinate of vector $\tilde{X}_n$ has the form

$$\lim_{n\to\infty} | x_{i,n} | \leqslant \Delta \sum_{k=1}^{m} | \check{h}_{ik} | , \qquad (281)$$

where $\check{h}_{ik}$ are elements of matrix $\check{H} = H^{-1}\tilde{H}$.

Let us consider the solution of the problem of synthesis of adaptive control for a special case of system (260) which corresponds to scalar difference equation (A.2) of the form

$$x_{n+1} = -a_1x_n - a_2x_{n-1} + b_1u_n + b_2u_{n-1} + f_n , \qquad (282)$$

where $a_1 = 1$, $L^T = (-a_1, -b_1, -b_2)$ is vector of constant but unknown parameters for which a priori estimate of the form (266) is given and restriction $b_1 > 0$ is imposed on the value of parameter b_1.

We shall select control u_n at each step from the condition of minimization of specific loss function of the form (261). In the special case being considered here, the function takes the form

$$\omega_{n+1} = | \varphi_n + \check{l}_1 z_{1,n} + \check{l}_2 z_{2,n} + \check{l}_3 z_{3,n} | , \qquad (283)$$

where

$$\varphi_n = (1 - a_1 b_2 b_1^{-1})(-a_1 x_n - x_{n-1} + b_2 u_{n-1}) - b_2 b_1^{-1} x_n +$$

$$+ b_1^{-1}(b_2^2 + b_1^2 - a_1 b_1 b_2) u_n + (1 - a_1 b_1^{-1} b_2) f_n ,$$

$$z_{1,n} = (a_1 - b_1^{-1} b_2) x_{n-1} + x_n - b_1 u_{n-1} ,$$

$$z_{2,n} = -x_{n-1} - b_1^{-1} b_2 x_n + b_2 u_{n-1} ,$$

$$z_{3,n} = b_1^{-1} b_2 x_{n-1} + (a_1 b_1^{-1} b_2 - 1) x_n + b_1^{-1} b_2^2 u_{n-1} .$$

Here we assume that the condition of controllability (264) for the system (282) which takes the form

$$b_2^2 + a_2 b_1^2 - a_1 b_1 b_2 \neq 0 , \tag{284}$$

in the case being considered here is fulfilled for true parameter values.

Disturbance f_n satisfies condition (268), i.e.

$$| f_n | \leqslant \delta = \text{const} \qquad \forall\, n \geqslant 0 . \tag{285}$$

Let us solve problem (269) for system (282), i.e. let us determine control sequence u_n from the solution of the problem

$$\min_{u_n} \max_{\substack{L \in \Omega_n \\ f_n \in f}} \{ \omega_{n+1} \} , \tag{286}$$

where ω_{n+1} is the loss function defined by expression (283) and Ω_n is the sequence of parameter vector estimates defined in accordance with recurrence expression (85). The observation equation used in constructing the sequence of estimates Ω_n is equation (282).

A numerical simulation of the problem being considered was carried out for the case when set Ω_0 is a parallelepiped in three-dimensional space {L} with vertices L_i , $i \in \overline{1,8}$. In this case, the sequence of values of disturbance f_n was generated by means of random number generator with the numbers uniformly distributed in the interval $[-\Delta, \Delta]$. True value of parameter vector $L = \overset{*}{L}$ was selected so as to make the controlled plant unstable and non-minimum-phase. An asymptotically stable reference system was selected. **Figs** 17 and 18 present the results of the simulation for $\overset{*}{L}^T = (-2.69; 1; -1.2)$ and $\Delta = 1$ with a priori estimates $-3 \leqslant a_1 \leqslant -2.5$, $0.5 \leqslant b_1 \leqslant 1.5$, $-1 \leqslant b_2 \leqslant 0.5$ when initial conditions in (282) have been selected as follows: $x_0 = \pm 5$, $x_{-1} = u_{n-1} = 0$. In this case the reference system parameters have been selected to be $\check{l}_1 = \check{l}_2 = \check{l}_3 = 0$, i.e. from the condition of the maximum speed of response of the closed-loop system. **Fig.** 17 presents the transient processes of the output coordinate x_n in the closed-loop control system and **Fig.** 18 shows the variation of control u_n .

As shown in the figures, diameter of the set $\delta(\Omega_n)$ is decreased from the initial value $\delta(\Omega_0) = 1.22$ to $\delta(\Omega_9) = 0.02$ already in 9 steps and the initial deviation of the output coordinate x_n is decreased to the values less than one. Similar results have been obtained also at other values of the plant and reference system parameters as well as at more rough a priori estimates for which inequality (233) holds.

Thus, the proposed method of constructing adaptive control systems is rather efficient in solving the problem of stabilizing linear controlled plants in the most general case, i.e. in case of unstable and non-minimum-phase plants. The dissipativity of a closed-loop system subjected to uncontrollable disturbances f_n is provided at a sufficiently high initial degree of uncertainty regarding the controlled plant parameters and a sufficiently high quality of the transient processes in a control system is provided.

Let us point out that no appreciable growth in the number of vertices and faces was observed in constructing the sequence of estimates. Besides, the process of identification itself terminated automatically at a rather small $\delta(\Omega_n)$.

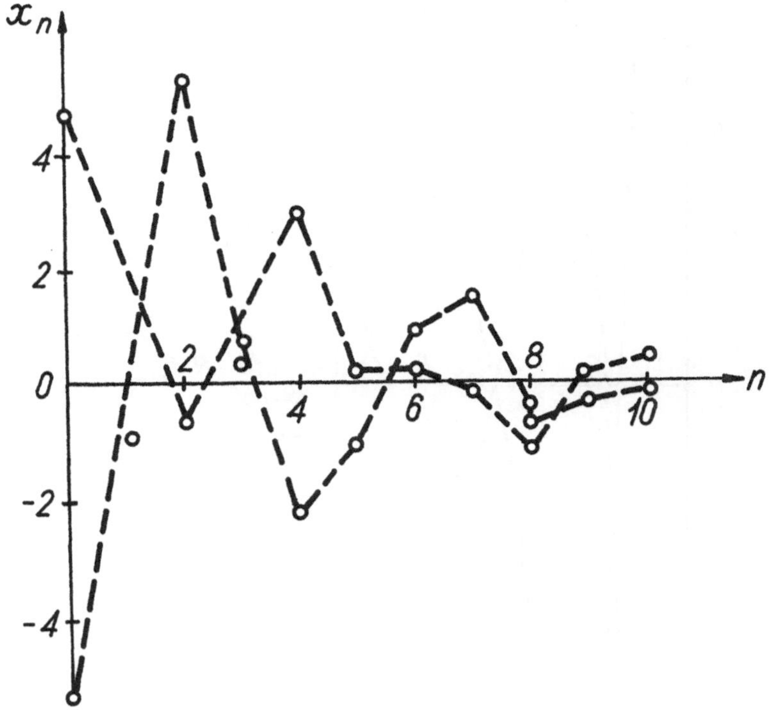

Fig. 17. Illustration of the process of adaptive control of a non-minimum-phase system

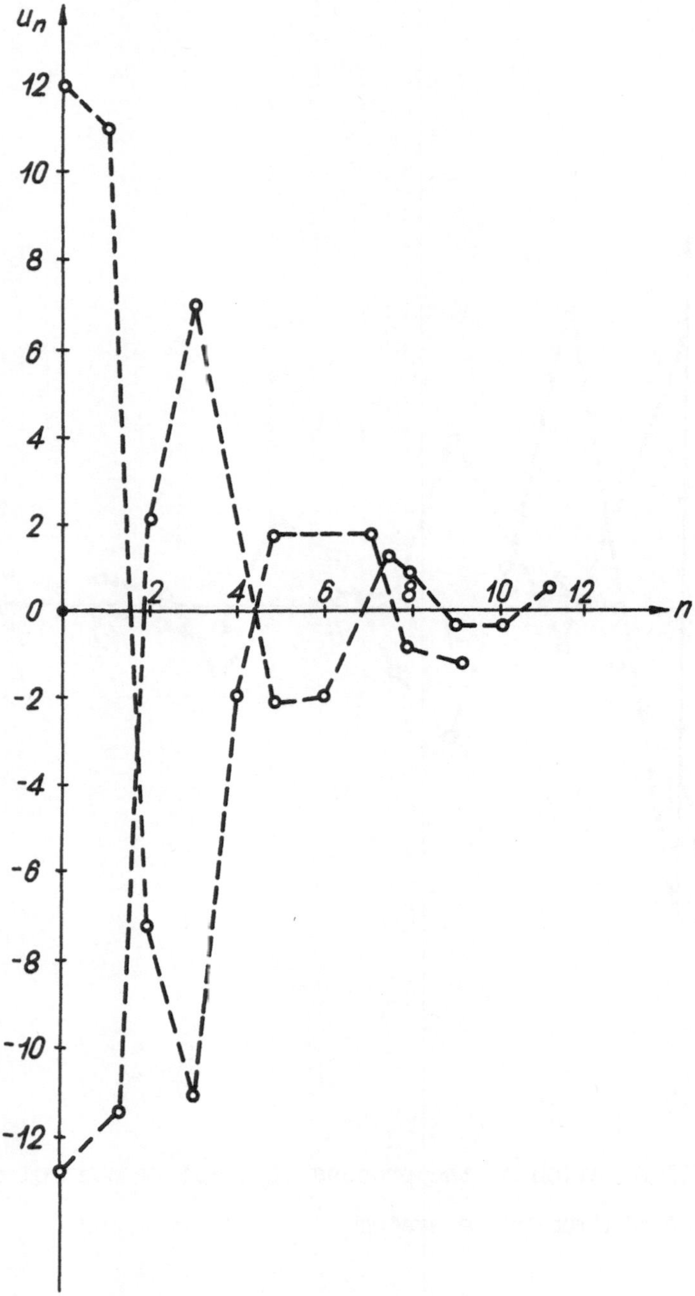

Fig. 18. Variation of control in the process of adaptive control of a non-minimum-phase system

2.4. Synthesis of Adaptive Optimal Stabilizing Systems for Nonstationary Plants

Let us consider the problem of control of nonstationary controlled plants whose parameters (all or only a part of them) vary in time by arbitrary laws. As above, we shall assume that the controlled plants are acted upon by uncontrollable and restricted in some sense external disturbances. In this case we shall use the same as above mathematical models of controlled plants which we shall describe here once more for convenience in presentation. Thus, the equations of the motion of controlled plants are assumed in the form

$$X_{n+1} = A(L_n)X_n + B(L_n)u_n + Cf_n , \quad n = 0, 1, 2, \ldots , \tag{287}$$

where

$$A(\cdot) = \left\| \begin{array}{c|c} 0 & I_{m-1} \\ \hline \multicolumn{2}{c}{\bar{L}_n^T} \end{array} \right\| , \qquad B(\cdot) = \left\| \begin{matrix} 0 \\ \vdots \\ 0 \\ b_n \end{matrix} \right\| , \qquad C(\cdot) = \left\| \begin{matrix} 0 \\ \vdots \\ 0 \\ 1 \end{matrix} \right\| . \tag{288}$$

Here $L_n^T = (\bar{L}_n^T, b_n)$ is $(m + 1)$-dimensional vector of time-varying parameters for which its a priori estimate for $n = 0$ is given

$$L_0 \in \varrho_0 , \tag{289}$$

where ϱ_0 is a given bounded convex set (polyhedron), f_n is an uncontrollable disturbance for which its a priori estimate is given

$$f_n \in f \qquad \forall\, n \geqslant 0 . \tag{290}$$

Like above in **Section 1.3.**, we shall consider only the case of restricted rate of variation of parameter vector L_n , i.e.

$$\| \Delta L_n = L_{n+1} - L_n \| \leqslant \tilde{\delta} = \text{const} \qquad \forall\, n \geqslant 0 . \tag{291}$$

In this case, the value of this restriction $\tilde{\delta}$ is a priori known to the system designer. Let us consider here the simplest case when restrictions

$$| \Delta l_{i,n} | \leqslant \tilde{\delta}_i = \text{const} \qquad \forall\, n \geqslant 0\,, \qquad i \in \overline{1,m+1} \tag{292}$$

are known for each component $\Delta l_{i,n} = l_{i,n+1} - l_{i,n}$ and when $f = \emptyset$ in (290).

Let the objective of control be like above the optimal stabilization and let to be assumed as the specific loss function the value of Lyapunov function of the first difference at the (n + +1)-th instant of time with the function being selected of the same class as (260) but with regard to the non-stationarity of the controlled plant, i.e. let us assume

$$v_n = X_n^T P_n X_n\,, \tag{293}$$

where

$$P_n = \left\| \begin{array}{c|c} \bar{P}_n & 0 \\ \hline 0 & p_{mm,m} \end{array} \right\|, \qquad \left. \begin{array}{l} \bar{P}_n^T = \bar{P}_n > 0\,, \\ p_{mm,n} > 0 \quad \forall\, n \geqslant 0\,. \end{array} \right\} \tag{294}$$

Then Δv_n calculated by virtue of equation (287) is equal to

$$\Delta v_n = p_{mm,n+1} |\bar{L}_n^T X_n + b_n u_n|^2 + \bar{X}_n^T \bar{P}_{n+1} \bar{X}_n - X_n^T P_n X_n\,, \tag{295}$$

where $\bar{X}_n$ is (m-1)-dimensional vector formed from vector X_n by crossing out its first component.

Since in (295) only the first term of this expression depends on control u_n then let us assume as loss function $\omega_n(\cdot)$ to be minimized at each n-th step the following expression

$$\omega_n(\cdot) = p_{mm,n+1} |\bar{L}_n^T X_n + b_n u_n|^2 \tag{296}$$

and we shall solve the optimal stabilizing problem for controlled plant (287)

$$\min_{u_n} \max_{L_n \in \mathfrak{L}} \{ \omega_n(\cdot) = p_{mm,n+1} | \bar{L}_n^T X_n + b_n u_n |^2 \} . \tag{297}$$

The procedure of construction of the sequence of estimates $\mathfrak{L}_n^{(n)}$ was described in detail above in **Section 1.3.** (ref. to **Theorem 4** and expressions **(109)-(112)**) and therefore there is no need to dwell on it. However, the peculiarities of the problem of parametric identification of nonstationary plants noted earlier require some more exact definition as applied to the class of nonstationary controlled plants. Because of this, let us introduce the required generalization of respective definitions of adaptability.

Definition 6. A system comprising a nonstationary controlled plant with a restricted rate of parameter vector variation and a control unit synthesizing control of the class

$$u_n = u(\mathfrak{X}_n, \mathfrak{L}_n^{(n)}, \mathfrak{f}, \tilde{\delta}) \tag{298}$$

from the feasible region Ω , where $\mathfrak{X}_n$ and $\mathfrak{L}_n^{(n)}$ are the sequences of guaranteed a posteriori estimates of the state and parameter vectors, respectively, with the procedure of their obtaining described in detail above, $\mathfrak{f}$ and $\tilde{\delta}$ are given non-empty sets and a constant (a priori estimates of disturbances f_n and of the rate of variation of parameter vector L_n , respectively) is called a weakly adaptive system if there exists such instant of time $k < N$ that the inequalities

$$\omega_n < \omega_{n_0} , \qquad \delta_n^L < \delta_{n_0}^L \tag{299}$$

are fulfilled for all $n \geq k$, where $n_0 \in \overline{0,k}$.

Definition 7. We shall call adaptive such a weakly adaptive system for which at

$$\delta(\mathfrak{f}) = 0 \tag{300}$$

conditions

$$\lim_{n\to\infty} \omega_n = 0 , \qquad \lim_{n\to\infty} \delta_n^L = 0 \tag{301}$$

are fulfilled.

Definition 8. We shall call robust such adaptive system for which at

$$\delta(f) \leqslant \delta_1 = \text{const} , \qquad \tilde{\delta} \leqslant \delta_2 = \text{const} \tag{302}$$

conditions

$$\overline{\lim_{n\to\infty}} \ \omega_n \leqslant \bar{\omega}(\delta_1,\delta_2) = \text{const} , \tag{303}$$

$$\overline{\lim_{n\to\infty}} \ \delta_n^L \leqslant \bar{\delta}(\delta_1,\delta_2) = \text{const} \tag{304}$$

are fulfilled at any initial conditions, arbitrary realizations of f_n (out of a given class) and feasible variations of parameter vector L_n .

Now let us refer to the definition of optimal control in problem (297). Since this problem coincides up to definitions with the problem (223), (224) already considered above, then it is clear that **Theorem** 7 which reduces the optimal control problem to the solution of the problem of the search for the unique root of equation (225) remains valid also in the case being considered here, i.e. optimal control for problem (297) is the only root of the equation

$$\varphi(u_n, n) = 0 , \tag{305}$$

where

$$\varphi(u_n,n) = \max_{L_n\in\Omega_n^{(n)}} \{ \bar{L}_n^T X_n + b_n u_n \} + \min_{L_n\in\Omega_n^{(n)}} \{ \bar{L}_n^T X_n + b_n u_n \} .$$

Having defined the method of generating control sequence u_n for system (287), let us dwell on the refinement of some peculiarities of the parameter identification procedure of this nonstationary dynamic system. It was pointed out above in the proof of **Theorem 8** that determinant of the system of linear equations **(230)** from whose solution the estimates of parameter vector are determined vanish if and only if control u_n is linear function of vectors X_n . It is easy to show that control u_n is a linear function of X_n only in the case when set $\Omega_n^{(n)}$ consists of only one point $L = \overset{*}{L}$. Thus, the "increase" of set $\Omega_n^{(n)}$ results in the fact that the respective optimal control differs to an increasing extent from linear control. Therefore, $\det \mathfrak{X}_{n-1}$ is not only not equal to zero for any finite set $\Omega_n^{(n)}$ containing not only one isolated point $L_n = \overset{*}{L}_n$ but also some its neighborhood, but moreover, if set $\Omega_n^{(n)}$ comprises some neighborhood of point $L_n = \overset{*}{L}_n$ in the form of a sphere with radius ρ , then the estimate of the form

$$| \det \mathfrak{X}_{n+1} | \geqslant \varepsilon(\rho) > 0 \qquad \forall\, n \geqslant m + 1 \tag{306}$$

takes place, where $\varepsilon(\rho)$ is some positive monotonically increasing function. But then it follows from **(306)** that

$$\| \mathfrak{X}_{n-1}^{-1} \| \leqslant \mu(\varepsilon) < \infty \qquad \forall\, n \geqslant m + 1 ,$$

where $\mu(\varepsilon)$ is some positive function of ε which provides the boundedness of estimates $\Omega_n^{(n)}$ being obtained by the method described above. In other words, optimal control determined from the solution of problem **(297)**, i.e. from the condition of the minimization only of specific loss function ω_n provide such a sequence of the results of the measurements along the trajectory of the motion of controlled plant **(287)** which does not prevent the solution of the identification problem. Thus, true is the following

Statement 6. A system comprising nonstationary plant **(287)** with given estimates of the initial values of parameter vector **(289)** and the rate of variation of its parameters **(291)** and with given control unit synthesizing control u_n from the solution of problem **(297)** where sets $\Omega_n^{(n)}$ are determined in accordance with expressions **(109)-(112)** is adaptive in terms of **Definition 7** .

Now, let us dwell on the problem of existence of the maximum admissible rates of parameter vector variation at which control determined from the solution of problem (297) still provides the asymptotic stability of the closed-loop control system.

First of all it is necessary to point to several reasons which automatically exclude the possibility to obtain in analytical form the exact estimates of the maximum admissible rates ΔL_n of parameter drift which still do not result in the loss of the system stability. It is well known that we fail in obtaining the necessary and sufficient conditions of asymptotic stability in the general case even for a linear nonstationary system. But optimal control in the system being considered are knowingly nonlinear functions of phase coordinates which can not be expressed in analytical form. Therefore our maximum hope in the case being considered is to prove the existence of asymptotically stable systems synthesized in accordance with the solution of problem (297) at some rather small rates of drift of system parameters. This fact is established by

Theorem 10. A control system consisting of a linear nonstationary plant (287) with the rate of parameter variation restricted by condition (291) and a control unit generating control sequence u_n from the solution of problems (297) with the use of the procedure of construction of sequence $\Omega_n^{(n)}$ of parameter vector estimates in the form of recurrence relationships (109)-(112) is asymptotically stable at a sufficiently small value of $\| \Delta L_n \|$,i.e. at a sufficiently small rate of parameter variation.

To prove the trooth of the **Theorem** , let us consider a knowingly non-optimal control

$$u_n = -\overset{*}{b}{}_n^{-1} (\overset{*}{\bar{L}})^T X_n \; . \tag{307}$$

After its substitution into (287) we obtain the equation of a closed-loop system

$$X_{n+1} = \tilde{A}_n X_n \; , \tag{308}$$

where

$$\tilde{A}_n = \overset{\circ}{A} + \delta\tilde{A}_n , \qquad \overset{\circ}{A} = \left\| \begin{array}{c|c} 0 & I_{m-1} \\ \hline \multicolumn{2}{c}{0} \end{array} \right\| , \qquad \delta\tilde{A}_n = \left\| \begin{array}{c|c} 0 & 0 \\ \hline \multicolumn{2}{c}{\delta\bar{L}_n^T} \end{array} \right\| .$$

In writing the expression for matrix $\delta\tilde{A}_n$, the following designation was used

$$\delta\bar{L}_n = \Delta\bar{L}_n - \frac{\Delta b_n}{\overset{*}{b}_n} \overset{*}{\bar{L}}_n , \tag{309}$$

since it follows from (287), (288) and (307) and relationship $L_{n+1} = \overset{*}{L}_n + \Delta L_n$ that

$$x_{m+1,n+1} = (\bar{L}_n - \frac{b_n}{\overset{*}{b}_n} \overset{*}{\bar{L}}_n)^T X_n = (\Delta\bar{L}_n - \frac{\Delta b_n}{\overset{*}{b}_n} \overset{*}{\bar{L}}_n)^T X_n = \delta\bar{L}_n^T X_n$$

Let us calculate Lyapunov function (293) along the trajectory of the motion of system (308)

$$v_{n+1} = X_n^T \tilde{A}_n^T P_{n+1} \tilde{A}_n X_n .$$

Let matrix P_n have the form

$$P_n = P = \mathbf{diag} \{ p_i \}_{i=1}^m ,$$

where $p_i = \sum_{j=1}^{i} q_j$, $q_j > 0$ are some numbers.

Then

$$\Delta v_n = -X_n^T (Q - \delta\tilde{A}_n P \delta\tilde{A}_n) x_n , \tag{310}$$

where $Q = \mathbf{diag} \{ q_i \}_{i=1}^m$ is a positive definite matrix. Then it follows from (309) and (310) that at sufficiently small $\| \Delta L_n \|$ (and, therefore, $\| \delta\tilde{A}_n \|$) matrix $R = Q - \delta\tilde{A}_n P \delta\tilde{A}_n$ is positive definite, i.e. $\Delta v_n < 0 \quad \forall X_n \neq 0$ which guarantees the asymptotic stability of system (287) with control (307). But if $\Delta v_n < 0$ for a knowingly non-optimal control then inequality $\Delta v_n < 0$ is automatically true for the optimal control determined from the solution of problem (298). Thus, the asymptotic stability of the synthesized control system at sufficiently small rates of drift of the system parameters is proved. It is natural that the proof of the asymptotic stability at sufficiently small rates of parameter drift obtained here has a qualitative nature since the estimates used here are rather rough in norm and do not allow to obtain exact values of the maximum allowable (on retention of the asymptotic stability of the system) parameter drift rates. There values seemingly can be obtained only in simulating the system being considered. We shall present below the results of such simulation which confirm the retention of the asymptotic stability of the synthesized system at a sufficiently high rate of parameter drift.

Let us consider now the same controlled plant (287) but subjected to the action of additive disturbance f_n limited in absolute value, i.e. let $\mathcal{f} \neq \emptyset$ in (290). The objective of control we shall assume as before the optimal stabilization with Lyapunov function being calculated along the trajectory of motion of system (287), i.e.

$$v_{n+1} = p_{mm,n+1} |X_n^T \bar{L}_n^T + b_n u_n + f_n|^2 + \tilde{X}_n^T \tilde{P}_{n+1} \tilde{X}_n \ . \tag{311}$$

Repeating the same reasoning which have been presented above in substantiating the choice of specific loss function (296) we obtain for the case being considered here instead of (296)

$$\omega_n(\cdot) = p_{mm,n+1} |\bar{L}_n^T X_n + b_n u_n + f_n|^2 \ , \tag{312}$$

and finally we shall seek for control from the solution of the problems

$$\min_{u_n} \max_{\substack{L_n \in \Omega_n^{(n)} \\ f_n \in \mathcal{f}}} \{ |\bar{L}_n^T X_n + b_n u_n + f_n| \} \ . \tag{313}$$

Since the problem being considered now is some "union" of the problems already solved above, then it is clear that we can use the results obtained earlier. Let us introduce the following designations

$$\hat{X}_n^T = (X_n^T, 1) , \quad \hat{L}_n^T = (\bar{L}_n^T, f_n) , \quad N_n^T = (L_n^T, f_n)$$

and then we reduce problem (313) with regard to what has been said to the problem

$$\min_{u_n} \max_{N_n \in \mathfrak{N}_n^{(n)}} \{ |\hat{L}_n^T \hat{X}_n + b_n u_n| \} , \tag{314}$$

where set $\mathfrak{N}_n^{(n)}$ is a Cartesian product of sets $\mathfrak{L}_n^{(n)}$ and $\mathfrak{f}$. The difference between the solution of problem (314) and the solution of problem (297) consists only in the necessity to determine sets $\mathfrak{N}_n^{(n)}$.

Let us consider now one more special case which is interesting from the point of view of application when disturbance acting on the nonstationary controlled plant (287) has a limited rate of variation in time and the value of this estimate is a priori known to the designer of the control system. Thus, let us assume that the time variation of disturbance f_n obeys the following difference equation

$$f_{n+1} = f_n + \Delta f_n \tag{315}$$

and a priori estimate is given for Δf_n

$$\Delta f_n \in \tilde{\mathfrak{f}} \qquad \forall\, n \geqslant 0 , \tag{316}$$

where set $\tilde{\mathfrak{f}}$ is defined by the restrictions

$$| \Delta f_n | \leqslant \nabla = \text{const} . \tag{317}$$

Only a priori estimate is given for f_0

$$f_0 \in \mathfrak{f}_0^{(0)} . \tag{318}$$

The objective of control we shall assume as before the minimization of the given specific loss function and therefore the problem of the synthesis of control has the form (313) as before with the only distinction that a priori estimate for disturbance f_n is given only at $n = 0$. It is required to construct the sequence of estimates $f_n^{(n)}$ for all $n > 0$ on the basis of equations (287), (288), (315)-(317) jointly with a priori estimate (318).

Let us show that this problem can be reduced to the preceding one. To do this, let us introduce the following designations

$$\hat{X}_n^T = (X_n^T, 1) , \quad N_n^T = (\bar{L}_n^T, b_n, f_n) .$$

Then we obtain from (287), (288) that the last equation of the system (287) can be written in the form

$$x_{m,n+1} = N_n^T \hat{X}_n$$

and therefore the procedure of the construction of the sequence of parameter vector estimates described above remains virtually the previous one only the generalized parameter vector N_n has now dimension greater by one than vector L_n .

Let us consider now an example illustrating the application of the identification and control determination procedures to the simplest nonstationary plant.

Let controlled plant (287) be given at $m = 1$

$$x_{n+1} = l_n x_n + b_n u_n , \quad x_0 = \overset{\circ}{x} , \; n = 0, 1, 2, \ldots . \tag{319}$$

Let a priori estimate $\Omega_0^{(0)}$ of vector $L_0^T = (l_0, b_0)$ be given by the system of inequalities

$$\underline{l}_0 \leq l_0 \leq \bar{l}_0 , \qquad \underline{b}_0 \leq b_0 \leq \bar{b}_0 , \tag{320}$$

where $\underline{l}_0$, $\bar{l}_0$ and $\underline{b}_0$, $\bar{b}_0$ are given numbers and $\underline{b}_0 > 0$. Let the true value of vector $\overset{*}{L}_0$ (unknown to the designer of the control system) be equal to

$$\overset{*}{L}{}_0^T = (2, 1.5) .$$

The rate of parameter drift is restricted by the inequalities

$$| \Delta l_n | \leqslant \Delta_1 , \qquad | \Delta b_n | \leqslant \Delta_2 . \tag{321}$$

Let us assume that the true values of the parameter drift rates (unknown to the designer of the control system) are

$$\Delta l_n = \Delta_1 , \quad \Delta b_n = \Delta_2 \qquad \forall\, n \geqslant 0 .$$

Let us assume that control u_n can take only one of the fixed values

$$u_n = u_n^{(i,j)} = c_i \text{ sign } j , \tag{322}$$

where $j = -1; 1$,

$$c = \begin{cases} \delta_1 i & \text{at} \quad i \in \overline{0,N} , \qquad \delta_1 = \text{const} , \\ \delta_2 (i - N) & \text{at} \quad i \in \overline{N_1+1,N} , \qquad \delta_2 = \text{const}. \end{cases}$$

Specific loss function is

$$\omega_n = px_{n+1}^2 + ru_n^2 = p(l_n x_n + b_n u_n)^2 + ru_n^2 \quad \forall\, n = 0 , \tag{323}$$

where $p = \text{const} > 0$, $r = \text{const} > 0$.

In the case been considered (with regard to expression (322)), problem (297) takes the form

$$\min_{\substack{i \in \overline{0,N} \\ j \in \overline{-1,1}}} \ \max_{L_n \in \Omega_n^{(n)}} \{ p(l_n x_n + b_n u_n^{(i,j)})^2 + r(u_n^{(i,j)})^2 \} , \tag{324}$$

where a priori set $\Omega_0^{(0)}$ is given by inequalities (320).

Let us consider the solution of this problem at the following numerical values of parameters

$$\delta_1 = 0.05 \,,\quad \delta_2 = 0.25 \,,\quad N_1 = 5 \,,\quad N = 11 \,,\quad r = p = 1 \,,$$

$$\overset{\circ}{x} = 1 \,,\quad \underline{l}_0 = 1 \,,\quad \bar{l}_0 = 3 \,,\quad \underline{b}_0 = 1 \,,\quad \bar{b}_0 = 2 \,.$$

As it was mentioned above, the problem

$$\max_{L_n \in \mathfrak{L}_n^{(n)}} \{ \omega_n = \omega(x_n, u_n, L_n) \}$$

is a linear programming problem and, therefore

$$\overset{\circ}{L} = \arg\max \{ \omega_n \}$$

belongs to one of the vertices $L_0^{(k)}$ of rectangle $\mathfrak{L}_0^{(0)}$. Therefore, the solution of problem (324) can be replaced by the solution of the problem

$$\min_{\substack{i \in \overline{0,N} \\ j \in \overline{-1,1}}} \max_{k \in \overline{1,4}} \{ (l_n^{(k)} x_n + b_n^{(k)} u_n^{(i,j)})^2 + (u_n^{(i,j)})^2 \} \,.$$

Solution of this problem at $n = 0$ (obtained in a computer by realizing the algorithm of the exhaustive search of variants whose number in the case being considered here is equal to $2kN = 88$) has the form

$$\overset{*}{u}_0 = -1.25 \,.$$

Changing controlled plant (319) under the action of this control from state x_0 to a new state $x_1 = 0.12$ we obtain the estimate

$$L_0 \in \tilde{\mathfrak{L}}_0^{(1)} = \{ L_0 \mid l_0 - 1.25 b_0 - 0.12 = 0 \} \,. \qquad (325)$$

In this case we obtain from (320) and (325) estimate $\tilde{\mathfrak{L}}_0^{(1)}$ for vector L_0 in the form of straight line segment $l_0 = 1.25 b_0 + 0.12$

limited by the points $l_1^{(1)} = 1.37$, $l_2^{(1)} = 1$ and $l_1^{(2)} = 2.62$, $l_2^{(2)} = 2$.

Carrying out the operation of summing up set $\bar{\Omega}$ (given by inequality (321)) and set $\tilde{\Omega}_0^{(1)}$, we obtain set $\tilde{\Omega}_1^{(1)}$ in the form of a convex polyhedron with vertices having the following coordinates

$$\left.\begin{array}{lll} l_1^{(1)} = 2.52 , & b_1^{(1)} = 2.1 , & l_1^{(2)} = 2.72 , \\ b_1^{(2)} = 2.1 , & l_1^{(3)} = 2.72 , & b_1^{(3)} = 1.9 , \\ l_1^{(4)} = 1.47 , & b_1^{(4)} = 0.9 , & l_1^{(5)} = 1.27 , \\ b_1^{(5)} = 0.9 , & l_1^{(6)} = 1.27 , & b_1^{(6)} = 1.1 . \end{array}\right\} \qquad (326)$$

For $n = 1$ problem (324) takes the form

$$\min_{\substack{i \in \overline{0,N} \\ j \in \overline{-1,1}}} \max_{k \in \overline{1,6}} \{ (l_1^{(k)} x_1 + b_1^{(k)} u_1^{(i,j)})^2 + (u_1^{(i,j)})^2 \} , \qquad (327)$$

where components $l_1^{(k)}$, $b_1^{(k)}$ of vectors $L_1^{(k)}$ are given by expression (326). Solution of this problem gives

$$\overset{*}{u}_1 = -0.15 , \quad (L_1)_{opt} = L_1^{(3)} , \quad (L_1)_{opt} = L_1^{(4)} .$$

The change over of controlled plant (319) under the action of this control to state $x_2 = 0.012$ gives the estimate

$$L_1 \in \tilde{\Omega}_1^{(2)} = \{ L_1 \mid 0.12 l_1 - 0.15 b_1 - 0.012 = 0 \} .$$

The intersection of set $\tilde{\Omega}_1^{(2)}$ with set $\Omega_1^{(1)}$ defines set $\Omega_1^{(2)}$ in the form of a straigth line segment whose ends have the following coordinates

$$l_2^{(1)} = 1.28 , \qquad b_2^{(1)} = 0.93 ,$$

$$l_2^{(2)} = 2.73 , \qquad b_2^{(2)} = 2.1 .$$

Then, summing up sets $\tilde{\Omega}$ and $\Omega_1^{(2)}$ in accordance with expression (107) we obtain new set $\Omega_2^{(2)}$.

The subsequent course of the solution of the problem being considered is already sufficiently obvious and we do not present here the detailed results of this solution at the following control steps for brevity sake. Graphs illustrating the variations of the controlled coordinate x_n and the values of $\delta(\Omega_n^{(n)})$ equal to the maximum distance between the vertices of set $\Omega_n^{(n)}$ are presented in **Fig. 19.** As it is seen from this figure, the asymptotic stability of the controlled plant with control determined from the solution of the sequence of problems (324) does not take place in the case being considered. This is the consequence of the fact that linear restrictions (322) are imposed on control u_n . As a result, the mode of auto-oscillations with a sufficiently small amplitude sets in the system.

The results of calculations presented in **Fig.** 19 confirm the claims presented above that the procedure for solving the problem of parameter estimation of a nonstationary plant used here provides a finite size of sets $\Omega_n^{(n)}$ at $n \to \infty$.

2.5. Synthesis of Adaptive Optimal Stabilizing Systems with Measurement Errors

Let us consider now a generalization of the problem of optimal stabilization of a plant with unknown parameters presented above for the case rather often encountered in practice when the state vector is measured with a limited noise. Thus, let the motion of the controlled plant be described by equation (202) as before and at the beginning we shall consider the case when $f = \emptyset$ to simplify the problem. Let us also assume that a priori estimate (204) is given for parameter vector L of system (202). Let the measuring unit be described by equation (119) as before which we shall write down here once more for convenience

$$Y_n = X_n + Z_n \ . \tag{328}$$

Here

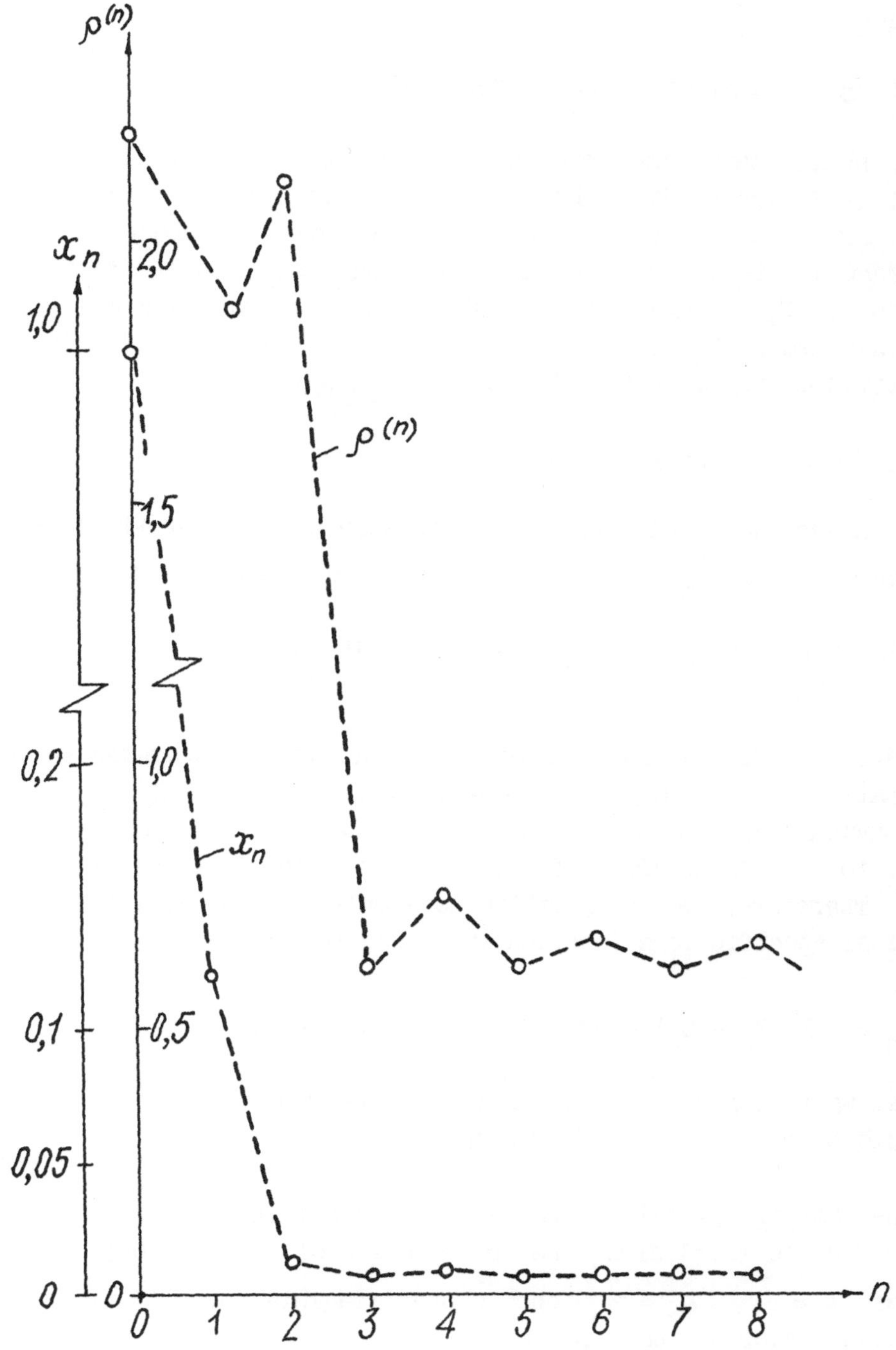

Fig. 19. Variation of controlled coordinate and diameter of set $\mathfrak{E}_n^n$ in adaptiv control of nonstationary system

$$Z_n \in \mathfrak{Z} \quad \forall\, n \geqslant 0\,, \tag{329}$$

where $\mathfrak{Z}$ is a given bounded convex set (polyhedron).

We shall assume the original control objective for the controlled plant (202) the minimization of specific loss function (208) at each n-th step. But since vector X_n is not measured directly in accordance with the conditions of the problem, then we determine the values of X_n from equation (328) in terms of the measured quantity Y_n and noise Z_n for which its a priori estimate (329) is given. Substitution of $X_n = Y_n - Z_n$ into (208) gives

$$\omega_n = \omega(Y_n, u_n, L, Z_n) \tag{330}$$

and, in accordance with what has been said earlier, we shall seek for optimal control $\overset{*}{u}_n$ from the solution of the sequence of problems

$$\min_{u_n} \max_{L \in \mathfrak{L}_n} \{ \omega(Y_n, u_n, L, Z_n) \}\,, \qquad n = 0,\, 1,\, \ldots\,. \tag{331}$$

Here $\mathfrak{L}_n$ is the sequence of the estimates of parameter vector L generated under conditions of the presence of noise Z_n by means of the recurrence procedure described in detail above in **Chapter 1** (ref. to expression (85)). There is no need to repeat its description here. Therefore, let us consider here only some peculiarities of the original specific loss function (208) has the form

$$\omega_n = |\, \bar{L}^T(Y_n - Z_n) + bu_n \,|\,. \tag{332}$$

Let us show now that the solution of problem (331) for function ω_n defined by the expression (332) gives

<u>Theorem 11.</u> Optimal control which provides the minimum of specific loss function (332) along the trajectories of system (202) at optimal values $L = \overset{*}{L}$, $Z_n = \overset{*}{Z}_n$ chosen over feasible sets $\mathfrak{L}_n$ and $\mathfrak{Z}$ is the unique root of the equation

$$\varphi(u_n) = 0\,, \tag{333}$$

where

$$\varphi(\cdot) = \max_{\substack{L\in\Omega_n \\ Z_n\in\mathfrak{Z}}} \{ \bar{L}^T(Y_n - Z_n) + bu_n \} +$$

$$+ \min_{\substack{L\in\Omega_n \\ Z_n\in\mathfrak{Z}}} \{ \bar{L}^T(Y_n - Z_n) + bu_n \} . \tag{334}$$

Proof. Recall that the case is being considered when $\infty > b > 0$. Let us introduce functions

$$\varphi_1(u_n) = \max_{\substack{L\in\Omega_n \\ Z_n\in\mathfrak{Z}}} \{ \tilde{\varphi}(Z_n,L,u_n) \} , \tag{335}$$

$$\varphi_2(u_n) = \min_{\substack{L\in\Omega_n \\ Z_n\in\mathfrak{Z}}} \{ \tilde{\varphi}(Z_n,L,u_n) \} , \tag{336}$$

where

$$\tilde{\varphi}(\cdot) = \bar{L}^T(Y_n - Z_n) + bu_n . \tag{337}$$

It follows from (337) that function $\tilde{\varphi}(\cdot)$ is linear in u_n with the restricted "slope" and because of this functions $\varphi_1(\cdot)$ and $\varphi_2(\cdot)$ are continuous and strictly monotonically increasing functions. In fact, function $\varphi_1(\cdot)$ is defined as the upper envelope of functions $\tilde{\varphi}(\overset{*}{Z}_n,\overset{*}{L},u_n)$ linear in u_n, where $\overset{*}{L} \in \Omega_n$, $\overset{*}{Z}_n \in \mathfrak{Z}$ take all their possible values out of Ω_n and $\mathfrak{Z}$ and coefficient $b > 0$ is limited. The "upper" envelope of this functions linear in u_n is a continuous and strictly monotonic function. The monotonicity and continuity of function $\varphi_2(\cdot)$ is proved similarly. Since function $\varphi(\cdot)$ is a sum of two continuous and strictly monotonically increasing functions $\varphi_1(\cdot)$ and $\varphi_2(\cdot)$ then it possesses itself the same properties and therefore equation (333) has one real root.

Now let us show that control u_n which is the root of equation (333) is optimal control. We shall prove this by contradiction. For example, instead of (333) at optimal control let

$$\varphi(\overset{*}{u}_n) > 0 \;. \qquad (338)$$

In this case it is obvious that

$$\max_{\substack{L \in \mathfrak{L}_n \\ Z_n \in \mathfrak{Z}}} \{ \omega(X_n, \overset{*}{u}_n, L, Z_n) \} = \varphi_1(\overset{*}{u}_n) > | \varphi_2(\overset{*}{u}_n) | \;.$$

However, in view of the consistency and monotonicity of functions $\varphi_1(\cdot)$ and $\varphi_2(\cdot)$, there exists such Δu_n that for

$$u_n = \overset{*}{u}_n + \Delta u_n \;,$$

where Δu_n is a sufficiently small number, the inequalities are true

$$\varphi_1(u_n) < \varphi_1(\overset{*}{u}_n) \quad \text{and} \quad \varphi_1(u_n) \geqslant | \varphi_2(u_n) | \;.$$

From this it follows that

$$\max_{\substack{L \in \mathfrak{L}_n \\ Z_n \in \mathfrak{Z}}} \{ \omega(X_n, u_n, L, Z_n) \} = \varphi_1(u_n) < \varphi_1(\overset{*}{u}_n)$$

and, therefore, suggestion (338) is false.

It can be shown in a similar way that the fulfillment of the inequality $\varphi(\overset{*}{u}_n) < 0$ contradicting (333) is impossible in optimal control.

A "union" in some way of **Theorems** 7 and 11 enables the solution to be obtained of the problem of the optimal control of a class of linear plants with unknown parameters also in the more general case when $\mathfrak{f} \neq \emptyset$ and $\mathfrak{Z} \neq \emptyset$ on which we shall not dwell here specifically. We shall also not dwell on the possibility of generalization of the obtained results of the solution of the optimal stabilization problem on the class of nonstationary systems with the restricted parameter variation rate which is rather obvious.

Let us consider an example illustrating the application of **Theorem 10** to determination of the optimal control of a simplest linear plant with unknown parameters with noise in phase coordinates measurement.

Let $m = 1$, then the equation of the controlled plant (202) has the form

$$x_{n+1} = lx_n + bu_n , \quad x_0 = \overset{0}{x} , \quad n = 0, 1, 2, \ldots ,$$

where l and b are unknown parameters about which it is known only that

$$l \in [1; 3] , \quad b \in [1; 2] ,$$

i.e. set ϱ_0 has the form of a rectangle in two-dimensional space of parameter vectors

$$\varrho^T = (l, b)$$

(ref. to **Fig. 20**). The measuring unit has the form (328), i.e.

$$y_n = x_n + z_n , \quad n = 0, 1, 2, \ldots ,$$

where it is known about disturbance z_n only that its values meet the inequality

$$\forall n \geqslant 0 \quad | z_n | \leqslant 0.2 .$$

In this case, specific loss function (332) has the form

$$\omega_{n+1} = | l(y_n - z_n) + bu_n | .$$

In simulating true motion of the plant, values z_n have been generated by means of a standard generator of random numbers (based on personal computer **IBM PC/AT**) uniformly distributed in the interval

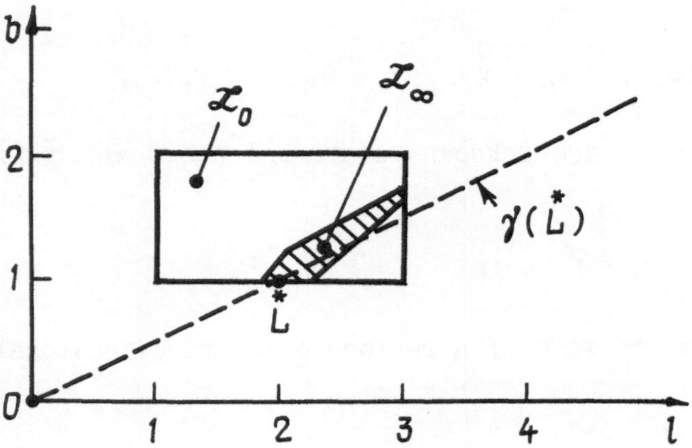

Fig. 20[a] Process of the set paramitric identification

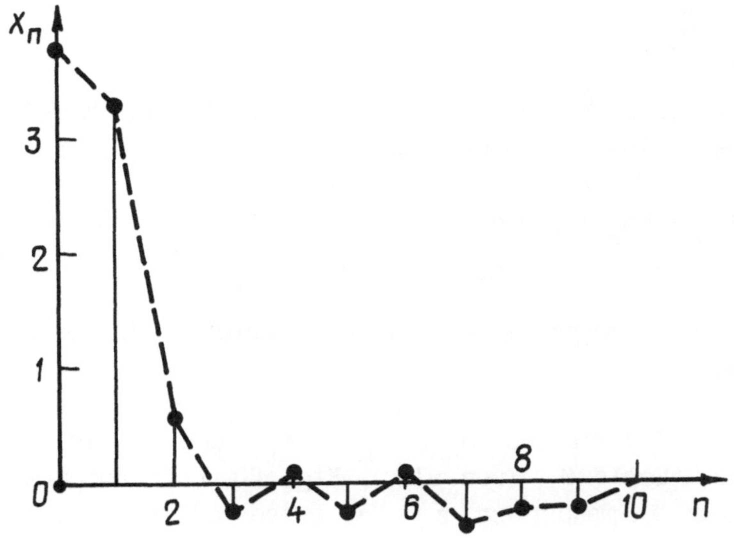

Fig. 20^b Process in the system with measurement errors

[-0.2; 0.2] and initial state was taken

$x_0 = 5$.

True values of parameters l and b have been selected, respectively:

$\overset{*}{l} = 2 , \quad \overset{*}{b} = 1$.

Optimal control $\overset{*}{u}_n$ was determined according to **Theorem** 10 from the solution of the nonlinear equation

$$\max_{L\in\Omega_n} \{ ly_n + bu_n \} + \min_{L\in\Omega_n} \{ ly_n + bu_n \} = 0 ,$$

where Ω_n is the sequence of parameter vector estimates mentioned in the theorem.

The values of variables z_n , u_n , x_n , and y_n at each step are presented in **Table 3** (see also **Fig.** 20). A comparatively high level of noise in measuring z_n has prevented the estimate of phase coordinate x_n obtained by means of a direct measurement from being refined at any step. Nevertheless, this circumstance has not hindered to refine substantially the a priori parameter vector Ω_0 . The best unimprovable estimate Ω_∞ obtained as a result of the set identification is shown cross-hatched in **Fig.** 20^a . As it is seen from the presented data, the presence of the disturbance z_n and the residual uncertainty of parameters naturally does not enable an asymptotically stable system to be obtained, but it provides for its dissipation. This means that the closed-loop control system synthesized in the present example is robust in terms of **Definition 5.**

With known parameters of the control plant and the absence of the measurement noise, the control objective, i.e. the obtaining of the minimum of function ω_{n+1} is achieved with control

$$u_n = cx_n , \quad \text{where} \quad c = \gamma(\overset{*}{L}) = -\frac{\overset{*}{l}}{\overset{*}{b}} .$$

Table 3

n	0	1	2	3	4	5	6	7	8	9
z_n	0.072	0.068	0.155	-0.080	0.129	-0.109	0.185	0.096	0.107	0.175
y_n	5.072	3.304	0.768	-0.293	0.252	-0.335	0.362	-0.230	-0.114	-0.052
x_n	5	3.237	0.613	-0.214	0.123	-0.226	0.176	-0.326	-0.221	-0.227
u_n	-6.763	-5.861	-1.440	0.550	-0.472	0.629	-0.679	0.431	0.214	0.099

Thus, optimal control in the present case can be determined not only at the single-point estimate $\wp$ containing point $\overset{*}{L}$ but also at the estimate in the form of a ray passing through the origin of coordinates and point L (ref. to dotted line in **Fig. 20[a]**). Therefore, the quality of the solution of the parameter identification problem in the case considered here should also be estimated not by means of such general estimates of the set $\wp_{\infty}$ as its diameter or volume, but by the dimensions of the cone inside which the set is fully located. The error in determining the estimate of $\gamma(\overset{*}{L})$ does not exceed $\sim 12\%$ in the case being considered.

2.6. Synthesis of Adaptive Control Systems Optimal for Functional

The solution of the problem of synthesis of adaptive control systems for linear plants with unknown parameters optimal with respect to some specific loss function was presented above. Particular attention has been given to a respective selection of the form of the specific loss function in order to add to the control system, along with the property of adaptability, such important qualitative properties as asymptotic stability (in the cases when it is attainable in principle) or dissipation (with the minimization of the dissipation domain). At the same time, for one or other reason it is necessary to solve a more general problem of synthesis of adaptive systems optimal with respect to a given functional. In view of this, let us consider in greater detail the problem statement of synthesis of adaptive control systems optimal with respect to a given aggregate functional as applied to the class of controlled plants with unknown parameters.

Let a class of controlled plants be given described by difference equations

$$X_{n+1} = \Phi(X_n, U_n, L, F_n, n) \,, \qquad X_0 = \overset{\circ}{X} \,, \quad n = 0, 1, 2, \dots \,. \tag{339}$$

Let a priori guaranteed estimates (204), (205) are given for the parameter vector of the plant and of the uncontrollable disturbances F_n and control U_n is selected from some feasible region Ω , i.e.

$$U_n \in \Omega \,. \tag{340}$$

Let also performance functional be given

$$J = \sum_{n=0}^{N-1} \tilde{\omega}(X_n, U_n, n) , \tag{341}$$

where $\tilde{\omega}(\cdot)$ is a given positive definite (or semi-definite) function.

Let us note here one qualitative peculiarity which is of principal importance in solving control synthesis problems for plants with unknown parameters as compared with the case when L_0 is a single-point set containing only one point $L = \overset{*}{L}$. The point is as follows. Let $\mathfrak{F} = \emptyset$, i.e. let disturbances F_n be absent. Then it is obvious that there exists such sequence of control $U_0, U_1, \ldots, U_{N-1}$, i.e. such vector $U^T = (U_0^T , U_1^T , \ldots , U_{N-1}^T)$ which minimizes functional (342) calculated along the trajectory of the motion of controlled plant (339) and this sequence retains its optimality property independently of any results of measurements along the trajectory of the motion. It is also obvious that this statement is no longer true in the case of control of a plant with unknown parameters with given their a priori estimate in the form of (204), since there is the possibility (considered above in detail) to refine the initial parameter vector estimates from the results of measurements along the trajectory of the controlled plant. In other words, if different instants of time n are equivalent in terms of decision making about the control of a controlled plant with known parameters, then it is not the case with these instants of time for a controlled plant with unknown parameters due to the possibility to carry out a continuous identification process. It is obvious that this circumstance should be taken into account in an explicit form in the procedure of synthesis of control for plants with unknown parameters.

First let us consider one of seemingly rather natural schemes of obtaining generally only suboptimal control*). Let us introduce along with performance functional (341) a sequence of functionals J_n on a "sliding" interval defined as

---

*)The reasons for which the control obtained below is generally only suboptimal will be described below.

$$J_n = \sum_{k=n}^{N-1} \tilde{\omega}(X_{k+1}, U_k, k) \ , \quad n = 0, 1, 2, \ldots, \ . \tag{342}$$

Then, in accordance with what has been said above, let us give the following definition of suboptimal control for the class of controlled plants (339) with the a priori parameter vector estimate in the form (204).

Definition 9. Let performance functional (342) be given for all values of state vector X_n and for each instant of time $n \geqslant 0$. The functional is calculated along the trajectory of the controlled plant (339) and the procedure of identification of parameter vector L is realized at each instant of time $n \leqslant 0$ based on the measurements taken along the trajectory of the plant (339). Then control

$$\overset{*}{U}_n = S_n \ (\tilde{U}_n)_{opt} \ , \tag{343}$$

where S_n is a block matrix of the form

$$S_n = \| \underbrace{I_l}_{l} \mid \underbrace{0_l \mid \ldots \mid 0_l}_{(N-1)\times l} \| \ , \tag{344}$$

I_l , 0_l are identity (unit) and zero matrices, respectively, of $(l\times l)$ dimension and $(\tilde{U}_n)^T_{opt} = (U^T_n, U^T_{n+1}, \ldots, U^T_{N-1})$ is determined from the solution of the problem

$$\min_{\tilde{U}_n \in \Omega} \ \max_{L \in \Omega_n} \ \max_{\tilde{F}_n \in \tilde{\mathfrak{F}}} \ \{ I_n \} \ , \tag{345}$$

where $(\tilde{F}_n)^T = (F^T_n, F^T_{n+1}, \ldots, F^T_{N-1})$, $\tilde{\mathfrak{F}} = \bigcup_{k=0}^{N-1} \mathfrak{F}_{n+k}$,

is called suboptimal control for plant (339) and performance functional (341).

Let us consider now the introduced procedure of calculation of suboptimal control values. Let problem (345) be solved for some values of X_n and n and respective feasible regions $\mathfrak{U}$, $\mathfrak{L}_n$ and $\widetilde{\mathfrak{F}}$ and the value of control $\overset{*}{U}_n$ determined in accordance with (343) be realized. In this case some set of values $L_{opt} \in \mathfrak{L}_n$ is determined. Since it is not known which value has taken disturbance vector F_n which can differ from $(\widetilde{F}_n)_{opt}$ in the general case, then for point X_{n+1} and for the instant of time $(n+1)$, The problem of the form (345) is to be solved anew based on the results of the executed identification, but now with the replacement of n by $(n+1)$ and the control U_{n+1} being found is realized. This procedure is repeated until at $n = N - 1$ the control at the last step is determined, i.e. $\overset{*}{U}_{N-1}$.

The solution of the suboptimal synthesis problem in the form of solution of a sequence of problems (345) predetermines a peculiar separation in time of the procedure of finding optimal control in point X_n , n and the procedure of obtaining a new a posteriori estimate L_{n+1} after the measurement of the results of the motion of the controlled plant under the effect of control $\overset{*}{U}_n$ found earlier. It is obvious that control determined in this way provides a better quality of control than if the problem (345) would be solved only one time at $n = 0$ and then the respective components of vector $(U_0)_{opt}$ would be realized sequentially. However the problem (345), even in the case when summation in (342) is carried out on a finite interval is a complex finite-dimensional nonlinear programming problem which is, generally speaking, practically solvable only for comparatively small values of N . This problem becomes practically unsolvable at a sufficiently large value of N . In this case, we have to restrict ourselves to the use of other algorithms more economical from the point of view of required computations at a sacrifice in our tendency to approach the optimal solution. Nevertheless, the problem of synthesis of control optimal with respect to the performance functional (341) remains extremely important at least since its solution makes it possible to evaluate the losses in the quality of control which are associated with the change from the optimal to suboptimal control of different kind dictated by our attempt to decrease substantially the volume of computations. Because of this, let us consider a solution of the problem of synthesis of control U_n, $n = \overline{0,N-1}$ optimal with respect to functional (341).

The presence of uncertainty regarding the values of vectors L and F_n for which only set estimates of the form (204) and (205) are given prohibits to use directly the motion equations (339) for the statement and solution of any control problems. In fact, even at $n = 1$, only a set estimate of the state of the system follows from (339), (204) and (205)

$$X_1 \in \mathfrak{X}_1 = \{ X_1 \mid X_1 = \Phi(X_0,U_0,L,F_0) \quad \forall L \in \mathfrak{L}_0 , F_0 \in \mathfrak{F} \} , \tag{346}$$

where set $\mathfrak{X}_1$ is not a single-point one.

Taking into account the possibility of the measurements of X_1 in the future, we obtain from (339) and (204) the forecasted estimate of parameter vector L and $n = 1$ in the form

$$L \in \check{\mathfrak{L}}_1, \ \check{\mathfrak{L}}_1 = \tilde{\mathfrak{L}}_1 \cap \mathfrak{L}_0, \ \tilde{\mathfrak{L}}_1 = \{ L \mid X_1 = \Phi(X_0,U_0,L,F_0) \quad \forall F_0 \in \mathfrak{F} \}, \tag{347}$$

where set $\tilde{\mathfrak{L}}_1$ depends on the value of X_1 not known in advance. In a similar way, at $n = 2$ we obtain the estimates

$$X_2 \in \mathfrak{X}_2 = \{ X_2 \mid X_2 = \Phi(X_1,U_1,L,F_1) \quad \forall L \in \tilde{\mathfrak{L}}_1, \ F_1 \in \mathfrak{F}, \ X_1 \in \mathfrak{X}_1 \}, \tag{348}$$

$$L \in \check{\mathfrak{L}}_2, \ \check{\mathfrak{L}}_2 = \tilde{\mathfrak{L}}_2 \cap \check{\mathfrak{L}}_1, \ \tilde{\mathfrak{L}}_2 = \{ L \mid X_2 = \Phi(X_1,U_1,L,F_1) \quad \forall F_1 \in \mathfrak{F} \}, \tag{349}$$

where set $\tilde{\mathfrak{L}}_2$ depends on unknown values of X_1 and X_2 given in advance which will be realized and then measured in the process of the motion. Thus, the mathematical model of the predicted motion of the system taking into account the uncertainty factors has the form of difference inclusions:

$$\begin{gathered} X_{n+1} \in \mathfrak{X}_{n+1} = \{ X_{n+1} \mid X_{n+1} = \Phi(X_n,U_n,L,F_n) \\ \forall L \in \check{\mathfrak{L}}_n , \ F_n \in \mathfrak{F} , \ X_n \in \mathfrak{X}_n \} , \quad n = 0, 1, 2, \ldots , \end{gathered} \tag{350}$$

and of the equations of evolution of sets $\check{\mathfrak{L}}_n$ which are the predicted estimates of parameter vector:

$$L \in \check{\Omega}_n , \quad \check{\Omega}_n = \tilde{\Omega}_n \cap \check{\Omega}_{n-1} , \quad \check{\Omega}_0 = \Omega_0 , \quad n = 1, 2, \ldots, \tag{351}$$

$$\tilde{\Omega}_n = \{ L \mid X_n = \Phi(X_{n-1}, U_{n-1}, L, F_{n-1}) \quad \forall F_{n-1} \in \mathfrak{F} \} , \tag{352}$$

where sets $\tilde{\Omega}_n$ depend on the values of X_n and X_{n-1} not known in advance.

In the motion of the system, in measuring the values of the phase coordinates vector X_n and in the realization of the set identification procedure the prediction of the further motion is carried out in accordance with the model (350)-(352) with the only difference that the value of X_n in (350) for any new initial instant of time $n = 0$ is considered to be known (i.e. $\mathfrak{X}_n = \mathfrak{X}_0$) and set Ω_0 is assumed to be equal to set $\check{\Omega}_0$ satisfying the set evolution equation

$$\check{\Omega}_n = \tilde{\Omega}_n \cap \check{\Omega}_{n-1} , \quad n = 1, 2, \ldots , \tag{353}$$

where set $\tilde{\Omega}_n$ is given by expression (352) but at the already known (measured) values of X_n and X_{n-1} .

The description of the system motion in the form of relationships (350)-(352) with due regard to (340) makes it possible to use the game approach in the statement of the optimal control synthesis problems, i.e. to reduce them to the solution of minimax problems on some sets and to obtain a control optimal in this sense.

The use of the plant motion model (350) and of the equations (351) of the evolution of the sets Ω_n makes it possible to search for the optimal control of the form

$$U_n = U (X_n, X_{n-1}, \ldots, X_0, U_{n-1}, U_{n-2}, \ldots, U_0, n), \quad n = \overline{0, N-1} \tag{354}$$

from the solution of the problem of minimization at the n-th step of functional (342) with regard to the a priori estimate of disturbances (205) and to the parameter vector estimate (204), obtained in accordance with the identification procedure (353). The expression "minimization" was used here in the general but not in the strictly mathematical sense. The problem of finding optimal U_n with regard to the uncertainty of the values forming functional (342) and the description of the plant (339) are written mathematically (in formalized way) as follows:

$$U_n = \arg \min_{U_n \in \mathfrak{U}} \max_{X_{n+1} \in \mathfrak{X}_{n+1}} \{ \tilde{\omega}(X_{n+1}, U_n, n) +$$

$$+ \min_{U_{n+1} \in \mathfrak{U}} \max_{X_{n+2} \in \mathfrak{X}_{n+2}} \{ \tilde{\omega}(X_{n+2}, U_{n+1}, n+1) + \ldots +$$

$$+ \min_{U_{N-1} \in \mathfrak{U}} \max_{X_N \in \mathfrak{X}_N} \{ \tilde{\omega}(X_N, U_{N-1}, N-1) \underbrace{\} \} \ldots \} \}}_{N-n+1} . \qquad (355)$$

The minimization on sets $\mathfrak{X}_{k+1}$ ($k = \overline{n,N-1}$) is carried out here with regard to relationships (350) relating these sets to unknown values of vector L which both enter directly under the symbol of function $\Phi(\cdot)$ and as a result of the use of the predicted estimates of $\check{\mathfrak{L}}_k$ from (351)-(352). The relationship between $\mathfrak{X}_{k+1}$ and disturbance vector F_k which is a part of $\Phi(\cdot)$ and F_{k-1} is also taken into account (via set $\check{\mathfrak{L}}_k$).

It follows from what has been said above that the problem being considered here belongs in accordance with the **M.A.Aiserman's** terminology [68] to the category of "pseudofunction" minimization problems since not only the values of the functions being minimized depend on the selection of vectors being varied, but also the sets on which the minimization procedure is carried out. In the present case, the solution of the optimization problem (355) is "aggravated"also by the fact that all minimization procedures are preserved also in solving the maximization problem.

When solving the problem of minimization of the expression (335) in U_k , $k \in [n+1, N-1]$, it is assumed that U_k are determined as functions analogous to (354) at $n = k$, $X_0 = X_n$ and $U_0 = U_n$. It should be pointed out in this case that all state vectors (except for X_n) and control vectors on which the predicted sets $\mathfrak{X}_{k+1}$ depend are unknown and they are in their turn the arguments of the function U_k

being sought. This circumstance extremely complicates the solution of problem (355) and results in the fact that the "cost" of the optimal solution is too high at any large value of N and the dimensionality of the system m and this forces the system designer to look for its suboptimal solutions as may be necessary. Thus, for example, to simplify the solution of the synthesis problem, it is possible to look for the control sequence not from the solution of the problem (355) but, as it was already described above, from the solution of the minimax problem for functional (342) in which all X_{k+1}, $k \in [n, N-1]$ are expressed in the long run in terms of X_n and sequences F_j and U_j, $j \in [k, N]$ where maximum is sought for in L and all F_k, $k \in [n, N-1]$ and minimum in all U_k, $k \in [n, N-1]$. In this case, only the first component of the control sequence, i.e. U_n is realized as the suboptimal solution at each n-th step. Of course, with this method of solution of the synthesis problem, predicted estimates (351)-(352) are not used but the possibility is preserved to take into account the interests of the following identification by imposing the additional restrictions on control vectors U_k, $k \in [n, N-1]$.

CHAPTER 3

ANALYSIS AND SYNTHESIS OF ROBUST CONTROL SYSTEMS

"... a mathematical solution can not be more exact than the approximations on which it is based".
Academician A.N.Krylov

3.1. Analysis and Synthesis of Control Systems under Uncertainty Conditions

The methods of solution of the problems of parameter estimates refinement and synthesis of optimal (in a sense stipulated above) control of plants under uncertainty conditions have been considered above in **Chapters** 1 and 2 . It was shown that under the conditions of available restricted uncontrollable disturbances the described procedures of construction of guaranteed estimates of controlled plants parameters provide the existence of some parameter estimates which are further unimprovable. However, this result should not be judged as a disadvantage intrinsic only in the method of solving parameter identification problems described here. The unimprovability of estimates, i.e. the existence of some "residual" unremovable uncertainty as regards the parameters is an objectively existing restriction preventing in the general case the estimates of the measured values with any preset accuracy from being obtained from the results of experiments. This objective restriction manifests itself to the same extent also in other identification procedures and it does not depend on the type of hypothesis on the nature of the uncontrollable disturbances (noise) whose presence excludes the possibility to obtain a trivial solution of the parameter identification problem. The final result of this process under real conditions always will be the obtaining only of some estimate of unknown parameters. The degree of closeness of this estimate to the true value of the quantity being measured depends on many factors: on the duration of the learning (identification) process, on real (not postulated in advance) properties of disturbances (noise), etc. In some cases the degree of closeness of these estimates to the true values of the parameters being identified can be so high that it is acceptable to identify the estimates with the real parameter values. But generally this can not be done and because of this a realistic approach to the problem statement of control synthesis inevitably being us to the necessity to solve the control synthesis problem under conditions of one or other uncertainty.

At present, there exists s difficulty countable number of publications devoted to the solution of the problem of robust control synthesis. A survey paper by **Siliak** [69] in which a rather comprehensive bibliography of the publications of the last few years is presented gives a rather comprehensive idea about the up-to-date state of the art in this field.

The solution of the problem of control synthesis of systems under uncertainty conditions is closely related to the solution of the problem of stability of these systems or, as it is used to say now, to the problem of robust stability.

We shall use below the following definition of this consept

Definition 9. If asymptotic overall stability takes place for the whole class of dynamic systems whose parameters belong to some given set then such class is called robust stable.

It is obvious that in the limiting case when the above set degenerates to one-point set, then the only dynamic system corresponding to this set is symply asymptotically stable as a whole.

Let us present here some results obtained in this field in the last few years [70].

3.2. Robust Stability of Continuous Dynamic Systems

Despite the fact that only discrete models of dynamic systems have been considered everywhere above in solving control and identification problems for the reasons which will becomeclear below, it is advisable to start the presentation of the proposed method of analysis of robust stability problem with the discussion of the robust stability of continuous dynamic systems.

The problem of robust stability of dynamic systems whose modern history originates from the pionering work by **V.L.Kharitonov** [71], has become in the last few years one of the most topical problems in control theory. The necessary and sufficient conditions of robust stability have been obtained in [71] for a class of continuous systems with the coefficient vector of characteristic equation given by interval estimates of its coordinates. In recent years the problem of robust stability of this class of systems has attracted widespread attention of researchers and many new results have been obtained here (e.g., ref. to [72]-[75]). Many papers are devoted to the investigation of the robust stability of discrete linear systems (e.g., ref. to [76]-[78] and to overview paper by **Jury** [79] which comprises a very comprehensive bibliography covering this problem), but nevertheless the attempts to obtain the necessary and sufficient conditions of the robust stability for a wide class of discrete dynamic systems have failed until very recently.

The robust stability problem for both continuous and discrete dynamic systems will be considered below from the same methodological positions.

Let the class of continuous systems be given

$$\dot{X} = AX , \quad X = \overset{\circ}{X} , \quad t \geq 0 , \tag{356}$$

where also like everywhere X is a m-dimensional state vector and A is a matrix with only the estimate known concerning its elements

$$A \in \Omega . \tag{357}$$

Here Ω is an arbitrary closed buonded set. Then one of possible statements (for the sake of definiteness we shall call it hereinafter "direct" statement) of the problem of robust stability of the class of systems (356), (357) is formulated as follows. It is required to find the necessary and (or) sufficient conditions of the existence of inclussion

$$\Omega \in \Omega_H , \tag{358}$$

where Ω_H is a set of Hurwitzean matrices, i.e. matrices for which real parts of all its eigenvalues are negative.

With this statement of the robust stability problem, it is necessary first of all to find set Ω_H . From the formal point of view, definition of the set is very simple.

To demonstrate this, let us consider the characteristic equation of system (356)

$$\det || A - \lambda I || = \psi(L, \lambda) = \sum_{i=0}^{m} l_i \lambda^i = 0 , \tag{359}$$

where $L = || l_i ||_{i=0}^{m} = F(A)$ is the coefficient vector of this equation.

In terms of set theory , the Hurwitz criterion is formulated as follows: if the inclusion takes place

$$L \in \mathfrak{H} , \tag{360}$$

$\mathfrak{H}$ is Hurwitz set determined in the form

$$\mathfrak{H} = \{ L \mid L > 0 , \quad \psi_j(L) > 0 , \quad j = \overline{1,k} \} , \tag{361}$$

then system (356) is asymptotically stable.

By inequality $L > 0$ in (361) we mean a system of coordinate-wise inequalities and $\psi_j(L)$ the well-known nonlinear functions Hurwitz diagonal matrices. Considering that $L = F(A)$, we obtain from (361) a definition of set Ω_H in the form

$$\mathfrak{A}_H = \{ A \mid F(A) > 0 , \quad \psi_j [F(A)] > 0 , \quad j = \overline{1,k} \} . \tag{362}$$

The structure of restrictions defining set $\mathfrak{A}_H$ is substantially more complex than the structure of restrictions for set $\mathfrak{H}$ since all restrictions in (362) are nonlinear and, what is more, there is a superposition of nonlinear functions in (362).

Because of this, determination of conditions under which inclusion (358) takes place is a rather complex problem which apparently has no constructive solution.

By virtue of what has been said, let us consider a different statement of the robust stability problem. Since only estimate (357) is given for matrix A in (356), then,therefore, there is also only estimate for vector L in (359)

$$L \in \mathfrak{L} = F(\mathfrak{A}) . \tag{363}$$

When abstracting from estimate (357), we can assume set $\mathfrak{L}$ to be given at least in some cases and then the problem of robust stability at given estimate

$$L \in \mathfrak{L} , \tag{364}$$

where $\mathfrak{L}$ is arbitrary closed convex set is reduced to the determination of the existence conditions of inclusion

$$\mathfrak{L} \subset \mathfrak{H} . \tag{365}$$

Let us note that if that matrix A in (365) has a canonical structure, i.e.

$$A = \left\| \begin{array}{c|c} C & I_{m-1} \\ \hline \multicolumn{2}{c}{A_m^T} \end{array} \right\| , \quad \text{where } A_m \in \mathfrak{A} ,$$

then set $\mathfrak{A}$ determines also the estimate of the normalized vector L, i.e. vector $\tilde{L}^T = (1, \bar{L}^T)$, where at $a_0 = \text{const} \neq 0$

$$\bar{L}^T = \left[\frac{-a_1}{a_0}, \frac{-a_2}{a_0}, \dots, \frac{-a_m}{a_0} \right],$$

since in this case

$$\bar{L} \in \bar{\mathfrak{L}} = -\overset{*}{\Omega} .$$

However, if the anther about the robust stability of matrix **A** of arbitrary from should be given to the researcher in terms of the initial estimate **(357)**, i.e. in the form of existence conditions of inclusion **(358)**, and since inverse operator $\mathbf{F}^{-1}$ does not exist for relationship $\Omega = \mathbf{F}^{-1}(\overset{*}{\mathfrak{L}})$, it is impossible to answer the question about the exitence of inclusion **(358)** on the basis of existence of inclusion **(364)**. In this case we have to give up the direct statement of the robust stability problem and we have to look for its different statements and different methods of its solution.

We shall cinsider below only the class of convex sets $\overset{*}{\mathfrak{L}}$ which can be described in analytical form

$$\mathfrak{L} = \{ \mathbf{L} \mid \varphi(\mathbf{L}, \mathbf{K}) \leqslant \rho = \mathbf{const} \} , \qquad \mathbf{(366)}$$

where $\varphi(\cdot)$ is a generally non-differentiable nonlinear function, **K** is parameter vector determining (jointly with $\varphi(\cdot)$) the geometry of set $\overset{*}{\mathfrak{L}}$. It is assumed that **K** is selected so that at $\rho \to 0$ degenerating set $\overset{*}{\mathfrak{L}}$ belongs to Hurwitz's set $\mathfrak{H}$.

From geometric point of view, the problem of determination of conditions at which inclusion **(360)** takes place consists in determination of that critical value $\overset{*}{\rho}$ of parameter ρ at which (with given $\varphi(\cdot)$ and **K**) the restrictive set is a set of the maximum volume inscribed into $\overset{*}{\mathfrak{H}}$, where $\overset{*}{\mathfrak{H}}$ is a closed Hurwitz's set. In other words, it is required to find the value of $\overset{*}{\rho}$ at which set $\overset{*}{\mathfrak{L}}(\overset{*}{\rho})$ is tangent to set $\overset{*}{\mathfrak{H}}$ in some point $\overset{*}{\mathbf{L}}$ (ref. to **Fig. 21**).

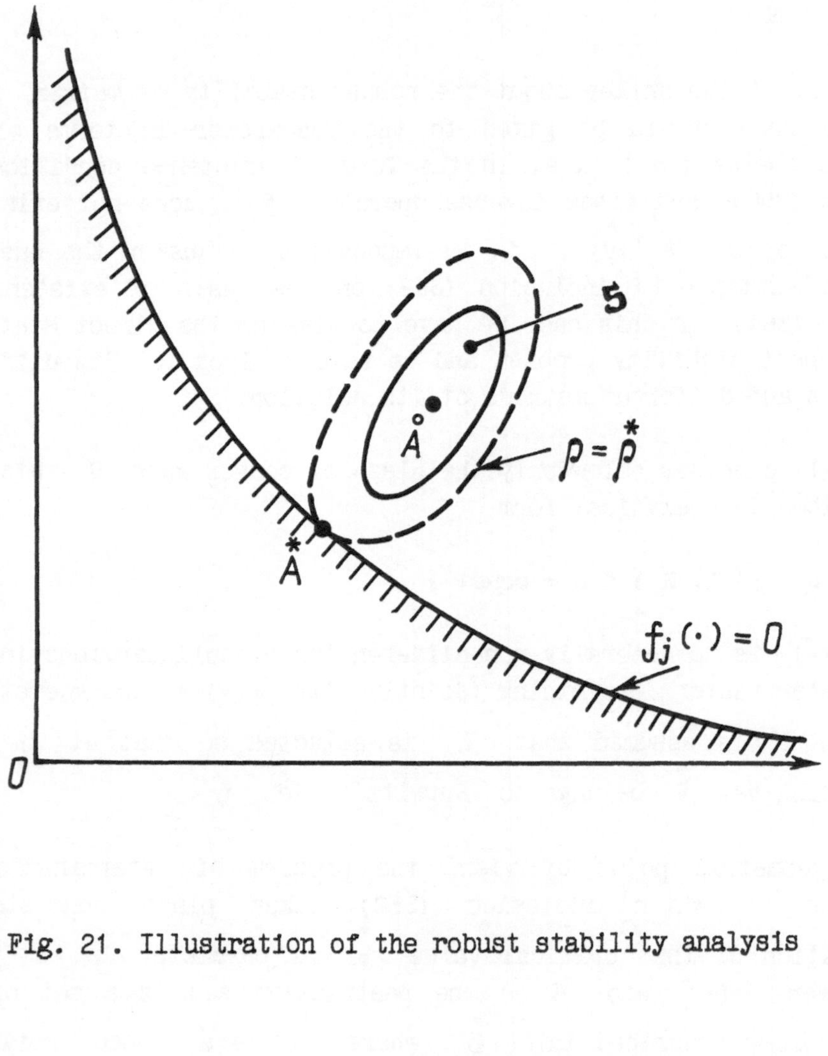

Fig. 21. Illustration of the robust stability analysis

Then it is obvious that true is

Statement 7. The fulfilment of inequality

$$\rho < \overset{*}{\rho} \tag{367}$$

is a necessary and sufficient condition of the robust stability of the class of systems (356),(366).

Once set $\bar{\mathfrak{H}}$ is introduced as a complement to set $\mathfrak{H}$

$$\bar{\mathfrak{H}} = R^{m+1} / \mathfrak{H} \tag{368}$$

the value of $\overset{*}{\rho}$ is determined from the solution of problem

$$\min_{L \in \bar{\mathfrak{H}}} \{ \varphi(A, K) \} , \tag{369}$$

since in this case

$$\overset{*}{\rho} = \varphi(\overset{*}{L}, K) , \tag{370}$$

where $\overset{*}{L} = \mathbf{argmin}\ \varphi(L, K)$.

The multiextremum nonlinear programming problem (369) (with difficulties in its solution being well-known) is equivalent to the search for minimum solution from solutions of the set of problems

$$\min_{L \geqslant 0} \{ \varphi(L, K) , \quad l_i = 0 \} , \quad i = \overline{0,m} , \tag{371}$$

$$\min_{L \geqslant 0} \{ \varphi(L, K) , \quad f_j(L) = 0 \} , \quad j = \overline{1,n} . \tag{372}$$

Let vectors $\bar{L}^i$, $i = \overline{0,m}$, and $\bar{L}^{m+j}$, $j = \overline{1,n}$ be the solutions of subproblems (371) and (372), respectively. Then $\overset{*}{\rho} = \min \{ \bar{\rho}^i , i = \overline{0,m+n} \}$, where $\bar{\rho}^i = \varphi(\bar{L}^i, K)$, $i = \overline{0,m+n}$.

The sense of restriction of problem (369) to the aggregate of subproblems (371), (372) consists in the fact that the minimization of objective function $\varphi(\cdot)$ in the latter is carried out on sets of structure substantially simpler than $\bar{\mathfrak{H}}$ which enables $\overset{*}{\rho}$ and $\overset{*}{L}$ to be found with much smaller expenditure of computations using quick-convergent minimization procedures whose application to the solution of problem (369) is is impossible or extremely inefficient. For instance, it is required in subproblem (371) to find a minimum of a convex function in a half-space with dimensionality m, in subproblem (372) - an a surface bounded by non-negative values of variables. The solution of the latter subproblem presents definite difficulties associated with non-convex nature of restrictions $f_i(L) = 0$. The problem is made easier only by constant signs of second partial derivatives $\partial^2 f_j / \partial l_i^2$ at $l_i \geq 0$, $i = \overline{0,m}$, at least for systems of not higher than tenth order.

The monotonicity property of function $f_j(\cdot)$ makes it possible to use for finding solutions of subproblems (372) the r-algorithm of non-smooth optimization [79] which has presented itself in a good light and the presence of the simplest restrictions $L \geq 0$ is efficiently used in its projective version [80]. In view of these reasons, the projective version of r-algorithm is used supplemented by the procedures of tough check of the stop with a random limited ejection (see below) to solve both the subproblems (371) and (372), the latter being reduced to the problem

$$\min_{L \geq 0} \{ \phi(L, K) + \alpha | f_j(L) | \}, \qquad j = \overline{1,n}, \tag{373}$$

by introducing a non-smooth penalty function, where $\alpha > 0$ is a penalty factor.

Let us consider the contensive sense of the projection of the r-algorithm on the example of solution of the problem $\min \{ \psi(x), x \geq 0 \}$ where $x \in R^N$, $\psi(\cdot)$ is a convex function. Intermediate point $\tilde{x}^{k+1} = x^k - r_k H_k g^k$ is plotted on some iteration k, where g^k is an element of $\partial_\Omega \psi(x)$, i.e. of the conditional subdifferential of function ψ on the set $\Omega = R^N_+$ of non-negative values of variables, H_k is the symmetrical positive definite matrix $N \times N$ (ref. to [79]),

k is a step multiplier. Vector g^k is constructed according to the rule: 1) if $x_1^k > 0$ then $g_1^k = [\ g(x^k)\ ]_1$; 2) if $x_1^k = 0$ and $-[\ g(x^k)\ ]_1 \leqslant 0$ then $g_1^k = 0$; 3) if $x_1^k = 0$ and $-[\ g(x^k)\ ]_1 > 0$ then $g_1^k = [\ g(x^k)\ ]_1$. We seek for the next point of the minimizimg sequence in the form $x^{k+1} = \mathrm{Proj}\ \tilde{x}^{k+1}\ /\ R_+^N$. Thus, variable x_1 turns out to be fixed on the boundary of the feasible region as long as condition 2 is satisfied. It was required to introduce some linear non-degenerate transformation of the original space (ref. to [80]) in order to exclude completely the variable x_1 from the process of optimization thus reducing substantially the computing efforts. The application of the transformation enables also information about variable x_1 inherent in matrix H_k to be restored if condition 3 is fulfilled on some iteration s , i.e. subgradient $g(x^s)$ prescribes a shift along x_1 inside the feasible region.

After fulfillment of the standard stopping criterion in solving problem (14) with the application of the projective version of r-algorithm, point $\tilde{L}_0$ will be found from ε-neighborhood of some local minimum. It might be well to check that $\| \tilde{L}_0 - \bar{L}^j \| \leqslant \varepsilon$. To do this, point $L \geqslant 0$ is selected in a random way sufficiently far from $\tilde{L}_0$ ($\| \tilde{L}_0 - L \| \leqslant E$, $E > \varepsilon$) , from which the descent is carried out with the use of the same minimization algorithm. The obtained approximate solution is compared with $\tilde{L}_0$, $\tilde{L}_1$ can be selected as $\bar{L}^i$, i.e. as the solution of problem (373), otherwise the procedure should be repeated for the point in which the objective function takes a smaller value.

The process with the random selection of the initial point can be repeated several times to be sure that the obtained approximate solutions fall within the ε-neighborhood of the global minimum. However, the practice has shown that there is no need to do this.

A special instrumental system **"Robust Stability"** [81] has been developed for the robust stability analysis. The system was intended for the solution of analysis problems in formulation (369) for the following classes of sets

$$1)\quad \Omega(\rho) = \{\ L \mid (\ L - \overset{\circ}{L}\ )^T Q\ (\ L - \overset{\circ}{L}\ ) \leqslant \rho\ \}\ , \qquad (374)$$

where $\mathfrak{L}(\rho)$ is (m+1)-dimensional ellipsoid, $\overset{\circ}{L} \in R^{m+1}$ is the coordinate vector of the center of the ellipsoid, $\overset{\circ}{L} \in \mathfrak{H}$, $Q = Q^T > 0$ is given matrix $(m+1) \times (m+1)$, ρ is given constant. In this case $\varphi(L, K) = (L - \overset{\circ}{L})^T Q(L - \overset{\circ}{L})$ and **k** is a vector consisting of all elements of $\overset{\circ}{L}$ and **Q** .

$$2)\quad \mathfrak{L}(\rho) = \{ L \mid \max_i \mid P_i^T (L - \overset{\circ}{L}) \mid \leqslant \rho \} , \tag{375}$$

where $\mathfrak{L}(\rho)$ is (m+1)-dimensional parallelepiped with arbitrary orientation; P_i is the i-th row of the matrix P defining the orientation of parallelepiped (375); $\varphi(L, K) = \max_i \mid P_i^T (L - \overset{\circ}{L}) \mid$; vector **k** comprises all elements of $\overset{\circ}{L}$ and P .

In a special case when P is a unitary matrix, $\mathfrak{L}(\cdot)$ is m-dimensional cube with aspects parallel to the coordinate axes. But if P is a diagonal non-unitary matrix then $\mathfrak{L}(\cdot)$ is m-dimensional rectangle with aspects parallel to coordinate axes (the case which has been considered by V.L.Kharitonov).

The consideration of these two classes of sets derives from the widespread use in recent years of the methods of obtaining guaranteed estimates of dynamic systems parameters in the form of ellipsoids and polyhedra.

The **"Robust Stability"** system is intended for operation under control of MS DOS version **3.3** and higher for PC type IBM AT with color monitor with arbitrary resolution. The system represents executable module with the size of **158** Kbyte [81]. In the present version of the system, dimension **m** of the problems is restricted by **10** . The indicated upper limit can be increased approximately by an order of magnitude for a PC with processor 80386 .

The time of solution the problems depends substantially on the availability of a coprocessor and it varies from several seconds to tens of minutes depending on the dimension of the problem (**m = 3-10**).

A number of robust stability problems was calculated by means of this system.

As an example, let us consider a class of systems of the fifth order with the vector of coefficients of the characteristic equation L equal to $L^T = (b_0, b_1, b_2, b_3, b_4, b_5)$. The Hurwitz's set $\mathfrak{H}$ with $m = 5$ is determined as follows

$$\mathfrak{H} = \{ L \mid L > 0 , \quad f_1(L) = b_1 b_2 - b_0 b_3 > 0 ,$$
$$f_2(L) = (b_1 b_2 - b_0 b_3)(b_3 b_4 - b_2 b_5) - (b_1 b_4 - b_0 b_5)^2 > 0 \} . \tag{376}$$

Let set $\mathfrak{Q}$ be given in the form of ellipsoid (374) for which $\overset{\circ}{L} = (1.5, 7, 12, 8, 2, 1)$, $Q = (H^{-1})^T H^{-1}$, where

$$H^{-1} = 16 \left\| \begin{array}{cccccc} 1 & 1 & 1 & 1 & 1 & 1 \\ 5 & 3 & 1 & -1 & -3 & -5 \\ 10 & 2 & -2 & -2 & 2 & 10 \\ 10 & -2 & -2 & 2 & 2 & -10 \\ 5 & -3 & 1 & 1 & -3 & 5 \\ 1 & -1 & 1 & -1 & 1 & -1 \end{array} \right\| . \tag{377}$$

The solution of problem (369) is vector $\overset{\circ}{L} = (1.49, 6.88, 11.87, 7.73, 1.83, 1.04)$ and, respectively, $\overset{*}{\rho} = 0.29$. In this case $\overset{*}{L} > 0$, $f_1(\overset{*}{L}) > 0$, $f_2(\overset{*}{L}) = 0$. Therefore, the point of tangency of ellipsoid (374) with the critical value of $\overset{*}{\rho}$ belongs to the nonlinear bound $f_2(L) = 0$ of Hurwitz's set $\mathfrak{H}$.

The robust stability for the given class of ellipsoids with the center in point $\overset{\circ}{L}$ and matrix Q takes place if and only if $\rho > 0.29$.

Let us consider as a second example a continuous system of the fifth order for which set $\overset{*}{\mathfrak{Q}}$ is given in the form of a family of parallelograms of the form (375) where $\overset{\circ}{L} = (0.5, 3.5, 8.5, 10, 6.5, 2)$ and matrix P has the form

$$P = \begin{Vmatrix} 1 & 0 & 0 & 0 & 0 & 2 \\ 2 & 0 & 0 & 0 & 0 & -1 \\ 0 & 1 & 0 & 0 & 2 & 0 \\ 0 & 2 & 0 & 0 & -1 & 0 \\ 0 & 0 & 1 & 2 & 0 & 0 \\ 0 & 0 & 2 & -1 & 0 & 0 \end{Vmatrix} .$$

Hurwitz's set is determined by expression (376) as before. As this takes place, the solution of problem (369) will be vector $\overset{*}{L}$ = (0, 3.5, 8.5, 10, 6.5, 1.83) and, respectively, $\overset{*}{\rho} = 0.83$. In this case, $f_1(\overset{*}{L}) > 0$, $f_2(\overset{*}{L}) > 0$, $l_0 = 0$.

In this example, point of tangency of the parallelogram from the family (375) with critical value $\overset{*}{\rho}$ belongs to the linear bound of Hurwitz's set $\mathfrak{H}$.

3.3. Robust Stability of Discrete Dynamic Systems

Despite a great quantity of papers dealing with the problem of the robust stability of discrete dynamic systems (e.g., ref. to [76]-[78]), this problem is developed much less than the similar problem for continuous systems. The truth of this statement follows at least from the fact that we failed to obtain a discrete analog of results obtained by V.L.Kharitonov even for such simplest sets as multidimensional cubes and rectangles with aspects parallel to coordinate axes. Let us show that the problem of the robust stability of discrete systems can be studied on the same methodological base as in the case of continuous systems.

Let the mathematical model of a dynamic system be given in the form

$$X_{n+1} = DX_n , \tag{378}$$

where like above X_n is the m-dimensional state vector, D is a matrix with dimension $m \times m$ concerning which it is known only that

$$D \in \mathfrak{D} . \tag{379}$$

Here $\mathfrak{D}$ is a given convex closed set.

The characteristic equation of system (378) has the form

$$\xi(R, z) = \det \| D_i - zI \| = \sum_{i=0}^{m} r_i z^{m-i} = 0 , \qquad (380)$$

where $R^T = (r_0, r_1, \ldots, r_m)$ is the coefficient vector of the equation. As to R , it is known only that

$$R \in \mathfrak{R} , \qquad (381)$$

where $\mathfrak{R}$ is a closed set obtained as a result of a functional transformation of the initial set $\overset{*}{D}$, i.e. $\mathfrak{R} = F(\mathfrak{D})$ whose form is determined by the structure of matrix D . As it was already mentioned above, if D is the accompanying matrix for (378), then the estimate of (379) determines also the estimate of the normalized vector $\tilde{R}^T$ = $(1, \bar{R}^T)$, where $\bar{R}^T = \left(\frac{r_1}{r_0}, \frac{r_2}{r_0}, \ldots, \frac{r_m}{r_0} \right)$. Then $\bar{R} \in \mathfrak{R} = - \mathfrak{D}$.

Hereinafter, let us assume that given is set $\mathfrak{R}$. It is obvious, that the robust stability of the system (378), (379) takes place if and only if the inclusion takes place

$$\mathfrak{R} \subset \mathfrak{S} , \qquad (382)$$

where $\mathfrak{S}$ is an open **Schur-Cohn** set.

As above in the analysis of continuous system, let us consider here only the case when set $\overset{*}{\mathfrak{R}}$ is described in analytical form.

$$\mathfrak{R} = \{ R \mid \varphi(R,G) \leqslant \rho = \text{const} \} . \qquad (383)$$

Here $\varphi(\cdot)$ is a generally non differentiable function, G is parameter vector determining (jointly with $\varphi(\cdot)$) the geometry of set $\mathfrak{R}$. It is assumed that G is selected so that at $\rho \to 0$ degenerating set $\mathfrak{R}$ belongs to **Schur-Cohn** set $\mathfrak{S}$.

From the geometrical point of view, the problem of determination of the conditions at which inclusion (382) takes place consists in the determination of the critical value $\overset{*}{\rho}$ of parameter ρ at which, with given $\varphi(\cdot)$ and G, the respective set is the set of the maximum volume inscribed into $\mathfrak{S}$ where $\mathfrak{S}$ is a closed Schur-Cohn set. In other words, it is required to find the value $\overset{*}{\rho}$ at which set $\mathfrak{R}$ is tangent to set $\mathfrak{R}$.

It is obvious that valid is

Statement 8. The fullfillment of inequality

$$\rho < \overset{*}{\rho} \tag{384}$$

is the necessary and sufficient condition of the robust stability of the class of discrete systems (378), (381).

Upon introducing set $\bar{\mathfrak{S}}$ as the complement to set $\mathfrak{S}$

$$\bar{\mathfrak{S}} = R^{m+1} / \mathfrak{S} , \tag{385}$$

the value $\overset{*}{\rho}$ is determined from the solution of problem

$$\min \{ \varphi(R,G) \} , \tag{386}$$

since in this case

$$\overset{*}{\rho} = \varphi(R,G) , \tag{387}$$

where $\overset{*}{R} = \mathbf{argmin}\, \varphi(R,G)$.

A more complex structute of restrictions in problem (380) as compared to problem (369) makes us to refuse from its direct solutions and to replace it with the equivalent statement but with a simpler sructure of restrictions. It is well known that the bilinear transformation of the form

$$z = \frac{p + 1}{p - 1} , \tag{388}$$

maps the unit circle of the complex plane into its left half-plane. Thus, the stability analysis of the original discrete system is reduced to the stability analysis of the respective continuous system by the change of variables (388). This change of variables corresponds to the linear transformation of the coefficients of characteristics equation **L** to the parameter vector **A** of the characteristic equation of the respective continuous system

$$\mathbf{A} = \mathbf{T}\mathbf{R}\,, \quad \det \mathbf{T} \neq 0\,, \tag{389}$$

where transformation matrix **T** is defined by the following system of equations (e.g., ref. to [64]-[67])

$$l_0 = \sum_{k=0}^{m} r_k\,,$$

$$l_\mu = \sum_{k=1}^{m} r_k \sum_{\nu=0}^{n} \binom{k}{\nu} \binom{n-k}{\mu-\nu} (-1)^\nu\,, \quad \mu = \overline{1,m-1}\,, \tag{390}$$

$$l_m = \sum_{k=0}^{m} (-1)^k r_k\,,$$

$\binom{k}{\mu} = \dfrac{k!}{\mu!(k-\mu)!}$ are binomial coefficients.

Once we have carried out linear transformation of the initial set $\mathfrak{R}$ by means of matrix T , we obtain the corresponding set $\overset{*}{\mathfrak{L}}$ in the space of coefficients of characteristics equation of the equivalent continuous system, i.e.

$$\mathfrak{L}(\rho) = \mathbf{T}\mathfrak{R}(\rho)$$

and, therefore, the problem of the robust stability of the discrete system is reduced as before to the check of existence of inclusion (365).

Thus, if function $\varphi(\cdot)$ in expression (383) has the form

$$\psi_1(\cdot) = (\overset{\circ}{R} - R)^T S (\overset{\circ}{R} - R)\,, \quad S^T = S > 0\,, \tag{391}$$

i.e. set $\overset{*}{\mathfrak{R}}$ is ellipsoid, then after the change of variables (388) in the space of parameters of the equivalent continuous system we obtain

$$\tilde{\varphi}_1(\cdot) = (\tilde{\overset{\circ}{L}} - L)^T\tilde{Q}(\tilde{\overset{\circ}{L}} - L), \tag{392}$$

where

$$\tilde{Q} = (T^{-1})^T S T^{-1}, \qquad \tilde{\overset{\circ}{L}} = T^{-1}\overset{\circ}{R}, \tag{393}$$

and for function

$$\psi_2(\cdot) = \| H(\overset{\circ}{R} - R) \|_{III} = \max_i | H_i^T(\overset{\circ}{R} - R) |, \tag{394}$$

determining set $\overset{*}{\mathfrak{R}}$ in the form of the m-dimensional polyhedron, after the substitution of (388) we obtain

$$\tilde{\psi}_2(\cdot) = \| \tilde{P}(\tilde{\overset{\circ}{L}} - L) \|_{III} = \max_i | \tilde{P}_i^T(\tilde{\overset{\circ}{L}} - L) |, \tag{395}$$

where P_i^T is the i-th row of the matrix

$\tilde{P} = HT^{-1}$, $\quad \tilde{\overset{\circ}{L}} = T^{-1}\overset{\circ}{R}$.

As an illustration of the described method of robust stability analysis, let us consider a discrete system of the fifth order.

Let the bound of set $\mathfrak{R}$ be given by the equation

$$\psi_1(\cdot) = (\overset{\circ}{R} - R)^T S(\overset{\circ}{R} - R) = \rho, \tag{396}$$

where $S = I$, $\overset{\circ}{R} \in \mathfrak{S}$.

In this case, by virtue of (396) and (393) we obtain that $\tilde{Q} = (T^{-1})^T T^{-1}$. Vector $\overset{\circ}{L} = T^{-1}\overset{\circ}{R}$ is assumed to be equal to $\tilde{\overset{\circ}{L}}{}^T = (1.5, 7, 12, 8, 2, 1)$ and, therefore, the problem of the analysis of the robust stability is reduced to the problem which has been already considered above.

Concluding the consideration of the problem of analysis of the robust stability of linear dynamic systems, let us note the following. The adequate language for the description and analysis of the motion of dynamic systems under uncertainty conditions is the language and the apparatus of differential and diference inclusions (e.g., ref. to [83]). Introducing designations

$$\mathfrak{X}(X) = \bigcup_{B\in\mathfrak{B}} (BX) \tag{397}$$

and

$$\mathfrak{X}_{n+1}(X_n) = \bigcup_{D\in\mathfrak{D}} (DX_n) , \tag{398}$$

we obtain from (356), (357) and (378), (379), respectively,

$$\overset{\circ}{X} \in \mathfrak{X}(X) \tag{399}$$

and

$$X_{n+1} \in \mathfrak{X}_{n+1} = \mathfrak{X}(X_n) . \tag{400}$$

Next, let us introduce such characteristics of set $\mathfrak{X}$ as its diameter

$$\delta_x = \delta(\mathfrak{X}) = \sup_{Y\in\mathfrak{X};\ Z\in\mathfrak{X}} \{ \| Y - Z \| \}$$

and its distance from the origin of coordinates

$$\rho_x = \rho(\mathfrak{X}) = \sup \{ \| X \| \} .$$

Then it follows from the obtained above necessary and sufficient conditions of the robust stability of dynamic systems that

$$\lim_{t\to\infty} \delta(\mathfrak{X}) = 0 , \qquad \lim_{t\to\infty} \rho(\mathfrak{X}) = 0$$

for differential inclusion and

$$\lim_{n\to\infty} \delta(\mathfrak{X}_n) = 0 , \qquad \lim_{n\to\infty} \rho(\mathfrak{X}_n) = 0$$

for difference inclusion. In this case, functions $\rho(\mathfrak{X}(t))$ and $\rho(\mathfrak{X}_n)$ take the part of Liapunov functions for systems (399) and (400), respectively (this is described in greater detail in [82]).

3.4. Synthesis of Optimal Robust Linear Control

Let the class of linear discrete systems (202) be given whose equation of motion at $\mathfrak{F} = \emptyset$ will be written here for convenience once more

$$X_{n+1} = AX_n + BU_n , \qquad X_0 = \overset{\circ}{X} , \quad n = 0, 1, 2, \dots , \tag{401}$$

where as above X_n is m-dimensional state vector, U_n is scalar control.

Matrix A $(m \times m)$ and m-dimensional vector B as above are assumed in canonical form, i.e.

$$A = \left\| \begin{array}{c|c} 0 & I_{m-1} \\ \hline \multicolumn{2}{c}{A_m^T} \end{array} \right\| , \qquad B = \left\| \begin{array}{c} 0 \\ \hline b \end{array} \right\| . \tag{402}$$

Let us assume that phase coordinate vector X_n is measured without niose, i.e. $Z = \emptyset$.

Let such rather robust unimprovable estimates be also given for system parameters A_m and b

$$A_m = \| +a_i \|_{i=1}^{m} \in \mathfrak{A}_m , \tag{403}$$

$$b \in \mathfrak{B} , \tag{404}$$

that we can not assume that $\sigma(\mathfrak{A}) \approx 0$ and $\sigma(\mathfrak{B}) \approx 0$.

Hereinafter, we shall assume estimates (403) and (404) to be interval ones, i.e. sets $\mathfrak{A}_m$ and $\mathfrak{B}$ are determined by the systems of inequalities

$$b = \{ b \mid \overset{\circ}{b} - \Delta \leqslant b \leqslant \overset{\circ}{b} + \Delta \} , \tag{405}$$

$$\Omega = \{ a_i \mid \overset{\circ}{a}_i - \Delta_i \leqslant a_i \leqslant \overset{\circ}{a}_i + \Delta_i , \quad i \in \overline{0,m-1} \} . \tag{406}$$

We shall consider the objective of the synthesis the determination of m-dimensional vector C in the equation of linear feedback

$$u_n = C^T X_n , \tag{407}$$

for system (401) which provides the required properties of the closed control system. The substitution of (407) into (401) gives

$$X_{n+1} = \tilde{A} X_n , \tag{408}$$

where

$$\tilde{A} = \left\| \begin{array}{c|c} 0 & I_{m-1} \\ \hline \multicolumn{2}{c}{\tilde{A}_m^T (\cdot)} \end{array} \right\| , \quad A_m(\cdot) = A_m + bC . \tag{409}$$

In this case the characteristic equation of system (408) has the form

$$\varphi(\lambda,\bar{R}) = \det (\tilde{A} - zI) = \sum_{j=0}^{m} \bar{r}_j z^{m-j} = 0 , \tag{410}$$

where

$$\bar{R} = \| \bar{r}_j \|_{j=1}^{m} , \qquad \bar{r}_j = -\tilde{a}_j . \tag{411}$$

Let us choose from the whole diversity of the methods of synthesis of linear control systems the reference model method and we shall tend to the properties of the model by varying vector C . As the model, let us assume a system characterized by the given (reference) disposition of roots $\overset{*}{Z}_i$, $i \in \overline{1,m}$ of its characteristic equation, i.e. a system characterized by vector $\overset{*}{Z} = \| \overset{*}{Z}_i \|_{i=1}^{m}$. Since the analytical dependence of vector of roots A of equation (410) on its variables C , A_m and b is absent and the solution of problem

$$\min \{ \| \overset{*}{Z} - Z(A_m, b, C) \| \} \tag{412}$$

is rather difficult, then, using Vieta formulas, let us determine from given vector $\overset{*}{Z}$ the corresponding values of coefficient vector $\overset{*}{R}_n$ of the characteristic equation of the reference system, i.e. we determine $\overset{*}{R} = R(\overset{*}{Z})$ and consider it to be given in what follows.

Since true values of A_m and b are not known and only their estimates (406) and (405), respectively, are given for them, then, therefore, the problem of the form (412) is ill-posed and it requires one or other extension of the definition. It seems to be natural, like above, to use such extension of the definition which provides a guaranteed result, i.e. let us formulate finally the synthesis problem in the form

$$\min_{C} \max_{A_m \subset \mathfrak{A}_m,\ b \in \mathfrak{b}} \{ \varphi(\cdot) = \| \overset{*}{A}_m - (A_m + bC) \| \} , \tag{413}$$

where $\| \cdot \|$ is the Euclidean norm of the vector.

Let us make one necessary reservation about the properties of set $\mathfrak{b}$, namely: let us make a natural assumption that it does not contain point $b = 0$. Otherwise we would have to deal with a problem which knowingly has no solution, i.e. with completely uncontrollable systems (401), (405), (406).

It is well known that minimax problems belong to the category of difficult problems, but in the present case there is analytical solution of problem (413) for the postulated above simplest structure of set $\mathfrak{A}_m$. This solution gives the following

<u>Theorem 12.</u> If sets $\mathfrak{b}$ and $\mathfrak{A}_m$ in problem (413) are defined by expressions (405) and (406) and $0 \notin \mathfrak{b}$, then the solution of this problem has the form

$$\overset{*}{C} = \operatorname{argmin} \max \{ \varphi(\cdot) \} = \overset{\circ}{b}{}^{-1} (\overset{*}{A}_m - \overset{\circ}{A}_m) , \tag{414}$$

where $\overset{\circ}{A}_m = \| \overset{\circ}{a}_i \|_{i=0}^{m-1}$.

Let us prove the theorem first for a special case with $\Delta = 0$, i.e. for the case of one-point set $\mathfrak{b}_m$. Let us introduce a new variable $\check{A}_m = \overset{*}{A}_m - A_m$ and set $\check{A}_m \in \check{\Omega}_m = \overset{*}{A}_m - \overset{*}{\Omega}_m$ corresponding to it. Hereinafter the algebraic sum of the sets is meant to be the Minkowski's sum. Let us introduce also notation $\overset{\circ}{C} = \overset{\circ}{b}C$. Then we can rewrite the problem in the form

$$\min_{\overset{\circ}{C}} \max_{\check{A}_m \in \check{\Omega}_m} \{ \| \check{A}_m - \overset{\circ}{C} \| \} .$$

It follows from the definition of the set $\overset{\circ}{\Omega}_m$ that it is a m-dimensional rectangle with the center in point $\check{A}^{\circ}_m = \overset{*}{A}_m - \overset{\circ}{A}_m$ (ref. to **Fig. 22**).

The necessity of the condition $\overset{\circ}{C} \in \check{\Omega}_m$ is obvious for the optimal vector $\overset{\circ}{C}$ in problem **(414)** which is also transferred to vector C with regard to the introduced notation. The sufficient condition of optimality in problem **(414)** is the fulfillment of the equality $\overset{\circ}{C} = \check{A}^{\circ}_m$ (which corresponds to **(413)** with regard to the introduced notation), which completes the proof of the **Theorem**.

The presence of the multipoint set $\mathfrak{b}$ at $\Delta \neq 0$ results in the fact that at any fixed vector $\overset{\circ}{C}$, vector $\tilde{C} = bC$ at $b \in \overset{\circ}{\mathfrak{b}}$ varies along vector $\overset{\circ}{C}$ within the limits of ray $\tilde{C} = [(\overset{\circ}{b} - \Delta)C; (\overset{\circ}{b} +\Delta)C]$ (ref. to **Fig. 22**).

The necessary condition of optimality in problem **(413)** at $\Delta \neq 0$ is as before the condition $\overset{\circ}{C} \in \check{\Omega}$. The sufficient condition as at $\Delta = 0$ is the fulfillment of **(416)** which proves the **Theorem** at $\Delta \neq 0$.

Although the possibility to generalize the obtained result to the case of the set $\overset{*}{\Omega}$ in the form of m-dimensional rectangle with arbitrary orientation about the coordinate axes is obvious, but the obtaining of the analytical solution of problem **(413)** is seemingly excluded. To use the result **(414)** in the latter case, the initial set of the parameters variation should be "coarsened" by inscribing it into the m-dimensional rectangle.

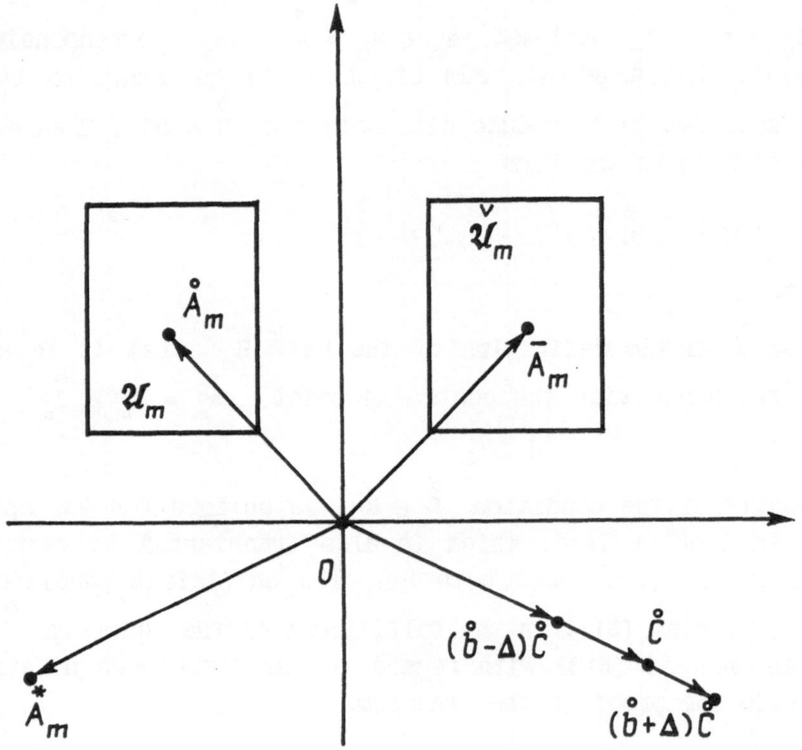

Fig. 22. Construction of set estimate of closed system parameters

The synthesis procedure is formally completed by defining vector **C** in the form of **(414)**. However, it is easy to show that even an optimal (in the sense mentioned above) chioce of the feedback parameters of course not only does not clear out the uncertainty of the parameters of the closed-loop control system (in view of relationships **(405)**, **(406)** and **(401)**, **(402)**) but also as a result of this does not guarantee the robust stability of the whole class of systems. This can be proved as follows: We obtain from **(406)** and **(401)**, **(402)** that

$$\tilde{\mathbf{A}}_m \in \tilde{\mathfrak{A}}_m , \tag{415}$$

where

$$\tilde{\mathbf{A}}_m = \mathfrak{A}_m + \mathfrak{b}\overset{*}{\mathbf{C}} , \tag{416}$$

and since set $\tilde{\mathfrak{A}}_m$ is the sum of two Minkowski's sets, then its diameter is greater than the diameter of set $\overset{*}{\mathfrak{A}}_m$.

Since the described procedure of optimal feedback synthesis pursued virtually only one objective: to achieve the desired location of the center of set $\tilde{\mathfrak{A}}_m$, then it is natural that among the elements of set $\tilde{\mathfrak{A}}_m$ will be found generally also such its elements which do not belong to **Schur-Cohn**'s set $\mathfrak{S}$,i.e. the robust stability of closed systems can also not take place.

It is possible to remove such undesirable result of the synthesis of control under uncertainty conditions in two ways.

The first of them which is virtually the "trial and error" method consists in the following: if inclusion **(382)** does not take place for vector **C** found from the solution of problem **(413)** (which can be found by means of the **"Robust Stability"** system) then the value of vector **C** changes with some step in direction to the normals in the point of tangency $\overset{*}{\mathbf{A}}$ the of **m**-dimensional rectangle of the maximum volume inscribed into **Schur-Cohn**'s set $\mathfrak{S}$ with the bound of this set. It is obvious that the realization of some procedure of the iterative correction of vector **C** can be generally required in order to fulfill condition **(382)**.

The second method of satisfaction of the robust stability conditions is based on the statement of control synthesis problem itself and on the account of these conditions in explicit form.

Let us consider a new statement of the problem of synthesis of optimal robust control (407) for the special case of system (401) when the uncertainty in the control channel is absent in this system. In other words, let us assume that set $\mathfrak{b}$ is single-point one comprising only one point $\overset{\circ}{b}$. Then we obtain from (406) and (416) that

$$\tilde{\mathfrak{U}}_m = \{ A \mid \tilde{a}_i(c_i) - \Delta_i \leqslant a_i \leqslant \tilde{a}_i(c_i) + \Delta_i , \quad i = \overline{1,m} \} , \qquad (417)$$

where $\tilde{a}_i(\cdot) = \overset{\circ}{a}_i + \overset{\circ}{b}c_i$.

Let us change from (417) to the other description of set $\tilde{\mathfrak{U}}_m$ in that its form which we have used above in analyzing robust stability, i.e.

$$\tilde{\mathfrak{U}}_m = \{ A \mid \varphi(A, \tilde{A}(C)) = \| \tilde{A}(C) - A \|_p \leqslant \rho \} , \qquad (418)$$

where $\| X \|_p = \max \mid c_i x_i \mid$, $\tilde{A}(C) = \| \tilde{a}_i(c_i) \|_{i=1}^m$, $\rho = c_i \Delta_i$.

<u>Statement 7.</u> Sets forth that robust stability takes place if

$$\gamma(C) = \min_{A \in \overline{\mathfrak{S}}} \{ \varphi [A, A(C)] \} - \rho > 0 . \qquad (419)$$

Therefore, we shall seek the optimal vector C for linear feedback (407) from the solution of the problem

$$\min_{C} \max_{A_m \in \mathfrak{U}} \{ \| \overset{*}{A} - (A_m + \overset{\circ}{b}C) \| \} \qquad (420)$$

provided that condition (419) is fulfilled.

Let us use the penalty functions method and reduce this problem to the equivalent problem of finding unconditional extremum

$$\min_{C} \max_{A_m \in \Omega} \{ \| \overset{*}{A} - (A_m + \overset{\circ}{b}C) \| + \alpha \max [0, -\gamma(C)] \} , \qquad (421)$$

where $\gamma(C)$ is given in form of (419) and $\alpha > 0$ is penalty coefficient.

It is seen from (421) and (419) that the second summand in (421) is equal to zero in the part of space $\{ A \}$ belonging to **Schur-Cohn**'s set $\mathfrak{S}$, i.e. where $\gamma(C) > 0$, therefore, vector C is determined as a mather of fact from the solution of problem (413).

Since the **Schur-Cohn**'s set $\mathfrak{S}$ is a bounded set, then the problem of synthesis of a robustly stable class of discrete systems has no solution in the general case at a sufficiently large diameter of set Ω and vector C obtained from the solution of problem (421) provides only a "uniform" in some way arrangement of set $\tilde{\Omega}$ with respect to $\mathfrak{S}$. This will be discussed in greater detail in considering a concrete example.

Let us illustrate the technique of synthesis of optimal robust control described above by the following example. Let the dimensionality of the system (401) be equal to 2 . Then matrix A and vector B are, respectively, equal to

$$A = \left\| \begin{matrix} 0 & 1 \\ a_{21} & a_{22} \end{matrix} \right\| , \qquad B = \left\| \begin{matrix} 0 \\ b \end{matrix} \right\| . \qquad (422)$$

Next, let the estimates of parameters b and $A_2^T = (a_{21}, a_{22})$ have the form

$$A_2^T \in \Omega_2 = \{ A_2 \mid \overset{\circ}{a}_{2i} - \Delta_i \leqslant a_{2i} \leqslant \overset{\circ}{a}_{2i} + \Delta_i , \quad i = \overline{1,r} \} , \qquad (423)$$

$$b \in \mathfrak{b} = \{ b \mid \overset{\circ}{b} - \Delta_b \leqslant b \leqslant \overset{\circ}{b} + \Delta_b \} \qquad (424)$$

with the following numerical values of constants in these estimates

$$\overset{\circ}{a}_{21} = 1.5, \quad \overset{\circ}{a}_{22} = 0.7, \quad \Delta_1 = 0.4, \quad \Delta_2 = 0.2, \quad \overset{\circ}{b} = 1, \quad \Delta_b = 0.1 .$$

Set Ω_2 corresponding to these values of the constants is constructed in **Fig. 23** . Then let the reference value $\overset{*}{Z} = 0$ and, therefore $\bar{R} = 0$.

To determine the optimal value of vector C in the equation of the linear feedback (407) which in the case being considered has the form

$$u_n = c^T x_n \text{ , where } c^T = (c_1, c_2) , \tag{425}$$

let us resort to **Theorem 12** . We obtain from (414) that

$$\overset{*}{C}{}^T = \overset{\circ}{b}{}^{-1} \overset{\circ}{A}_2 , \qquad \overset{\circ}{A}_2{}^T = (\overset{\circ}{a}_{21}, \overset{\circ}{a}_{22}) . \tag{426}$$

With this value of C the matrix of the closed system (408) has the form

$$\tilde{A} = \left\| \begin{array}{c|c} 0 & 1 \\ \hline \multicolumn{2}{c}{\tilde{A}_2^T} \end{array} \right\| , \qquad \text{where } \tilde{A}_2 = A_2 - \overset{\circ}{b}{}^{-1} \overset{\circ}{A}_2 , \tag{427}$$

In view of (426) and (423), (424), we obtain the estimate for vector $\tilde{A}_2$

$$\tilde{A}_2 \in \tilde{\Omega}_2 = \Omega_2 - \frac{1}{\overset{\circ}{b}} \overset{\circ}{A}_2 b . \tag{428}$$

Since vector C has been found from the solution of problem (413) which in the general case, as it was mentioned above, may not provide the robust stability of the synthesized system, then the necessary stage of the construction of a closed control system is the stage of the check of its robust stability.

Characteristic equation for matrix $\tilde{A}$ of the closed system has the form

$$\det \{ \tilde{A} - zI \} = z^2 - \tilde{a}_{22} z - \tilde{a}_{21} = 0 ,$$

which we shall rewrite in accordance with the notations introduced above (ref. to (411)) in the form

$$\bar{r}_0 z^2 + \bar{r}_1 z + \bar{r}_2 = 0 , \tag{429}$$

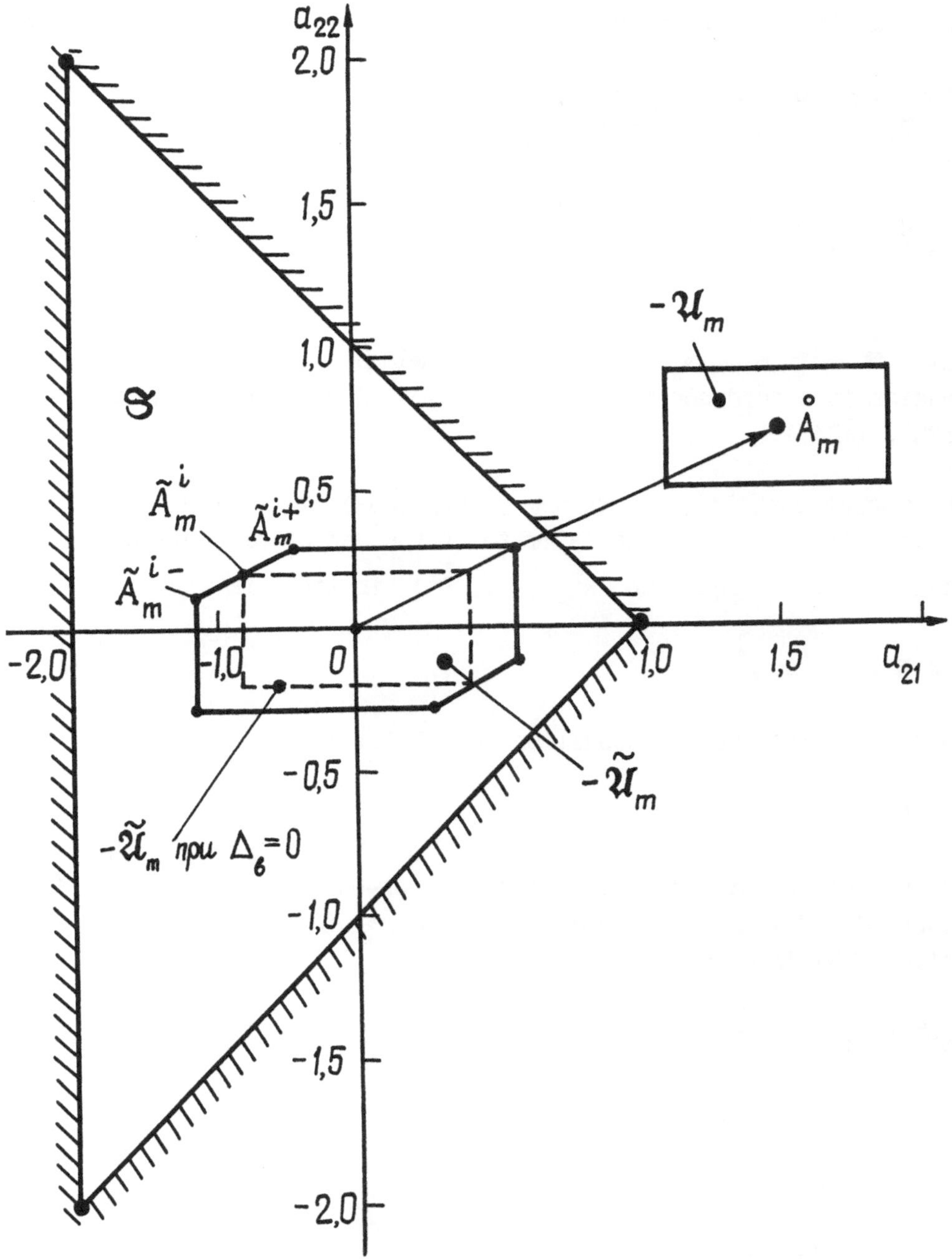

Fig. 23. Illustration of the robust stability analysis of a closed system

where $\bar{r}_0 = 1$, $\bar{r}_1 = -a_{22}$, $\bar{r}_2 = -a_{21}$. (430)

According to (430) and (428), (423), (424), we obtain for vector $\bar{R}^T = (\bar{r}_1, \bar{r}_2)$ only its estimate

$$\mathfrak{L} \in \bar{\mathfrak{R}} = -\tilde{\mathfrak{A}}_2 = -\mathfrak{A}_2 + \frac{1}{\overset{\circ}{\mathfrak{b}}} \overset{\circ}{A}_2 \mathfrak{b} . \quad (431)$$

Set $\bar{\mathfrak{R}} = \bar{\mathfrak{L}} = -\tilde{A}_m$ was constructed in **Fig. 23** from this expression in accordance with the estimates of sets $\mathfrak{A}_2$ and $\mathfrak{b}$ (ref. to (423), (424)).

The robust stability of system (401), (402), (425) with matrix $\tilde{A}$ (ref. to (427)) and with the estimate of the vector of coefficients of its characteristic equation $\tilde{R}^T = (1, \bar{R}^T)$ in the form

$$R \in \mathfrak{R} = 1 \times \bar{\mathfrak{R}} , \quad (432)$$

takes place only when the condition is fulfilled

$$\mathfrak{R} \subset \mathfrak{S} . \quad (433)$$

In the case being considered here, **Schur-Cohn** set $\mathfrak{S}$ determined by means of bilinear transformation (388) transforming characteristic equation (429) into equation

$$a_0 p^2 + a_1 p + a_2 = 0 , \quad (434)$$

where

$$\begin{aligned} a_0 &= r_0 - r_2 = 1 + a_{21} , \\ a_1 &= r_0 - r_1 + r_2 = 1 + a_{22} - a_{21} , \quad (435) \\ a_2 &= r_0 + r_1 + r_2 = 1 - a_{22} - a_{21} , \end{aligned}$$

has the form

$$\mathfrak{S} = \{\, L \mid l_0 - l_2 > 0, \quad l_0 - l_1 + l_2 > 0, \quad l_0 + l_1 + l_2 > 0 \,\} . \tag{436}$$

This set $\mathfrak{S}$ is constructed in **Fig. 23** by expressions **(435)**, **(436)**. It is seen from **Fig. 23** that inclusion **(433)** takes place and therefore the synthesized system is robustly stable.

The fulfillment of the robust stability conditions in the case considered here can be checked also without making recourse to the procedure of the graphical construction of set $\mathfrak{S}$ and to the subsequent "visual" analysis of the relative arrangement of sets $\bar{\mathfrak{L}}$ and $\mathfrak{S}$. Set $\bar{\mathfrak{L}}_2$ in fact can be described in the following form (ref. to **Fig. 23**)

$$\bar{\mathfrak{L}} = \mathrm{conv}\,\{\, \bar{L}_i^+,\ \bar{L}_i^- \,\} , \quad i = \overline{1,4} , \tag{437}$$

where

$$\bar{R}_1^+ = \left\| \begin{matrix} 0.57 \\ 0.27 \end{matrix} \right\| , \quad \bar{R}_2^+ = \left\| \begin{matrix} 0.57 \\ -0.13 \end{matrix} \right\| , \quad \bar{R}_3^+ = \left\| \begin{matrix} -0.23 \\ -0.13 \end{matrix} \right\| ,$$

$$\bar{R}_4^+ = \left\| \begin{matrix} -0.23 \\ 0.27 \end{matrix} \right\| , \quad \bar{R}_1^- = \left\| \begin{matrix} 0.23 \\ 0.13 \end{matrix} \right\| , \quad \bar{R}_2^- = \left\| \begin{matrix} 0.23 \\ -0.27 \end{matrix} \right\| ,$$

$$\bar{R}_3^- = \left\| \begin{matrix} -0.57 \\ -0.13 \end{matrix} \right\| , \quad \bar{R}_4^- = \left\| \begin{matrix} -0.57 \\ 0.23 \end{matrix} \right\| .$$

Further, since sets $\mathfrak{S}$ and $\bar{\mathfrak{R}}$ are convex then the check of existence of inclusion **(433)** is reduced to the check of the system of inequalities

$$S_j^T R > 0 , \quad j = \overline{1,3} , \tag{438}$$

where $S_1^T = (\,1,\ 0,\ -1\,)$, $S_2^T = (\,1,\ -1,\ 1\,)$, $S_3^T = (\,1,\ 1,\ 1\,)$ only for the vertices $\bar{R}_i^{\pm}$ of polyhedron $\bar{\mathfrak{R}}$, i.e. to the check of the system of inequalities

$$S_j^T R_i^+ > 0 , \quad S_j^T R_i^- > 0 , \quad j = \overline{1,3} , \quad i = \overline{1,4} , \tag{439}$$

where $(R_i^{\pm})^T = [\ 1,\ (\bar{R}_i^{\pm})^T\ ]$.

This system of inequalities is fulfilled for the example being considered.

Since set $\mathfrak{S}$ is bounded as it was already mentioned above, then it is obvious that the robust stability of the closed system is achieved only with the fulfillment of definite relationships between geometric characteristics of sets $\mathfrak{S}$, $\mathfrak{A}_2$ and $\mathfrak{b}$, "generating" set $\tilde{\mathfrak{A}}_2$. Specifically, with the arbitrary increase of the size of these sets, i.e. with the increase of constants Δ_1 , Δ_2 and Δ_b the robust stability is unattainable in principle.

3.5. Synthesis of Optimal Robust Control with Restricted Noise

Let us consider the problem of synthesis of robust control of a class of discrete systems in the presence of restricted noise, i.e. let us consider the problem of synthesis for system (202) whose equation of motion we write here down once more for convenience in presentation

$$X_{n+1} = AX_n + Bu_n + Cf_n . \tag{440}$$

In this equation, all notations preserve their initial meaning and we shall only recall that matrix A and vectors B and C have canonical structure (ref. to (203)) and that only a priori estimate (204) is given for parameter vector L of the system, i.e.

$$L^T = (\ A_m^T,\ b) \in \mathfrak{L} . \tag{441}$$

Let us consider below the case when set $\mathfrak{L}$ is the result of the Cartesian product of independent estimates

$$A_m \in \mathfrak{A} , \tag{442}$$

$$b \in \mathfrak{b} , \tag{443}$$

i.e.

$$\mathfrak{C} = \mathfrak{A} \times \mathfrak{b} . \qquad (444)$$

Here $\mathfrak{A}$ and $\mathfrak{b}$ are given closed convex sets. To exclude from the consideration the case of the complete uncontrollability of the class of systems **(440)**, let us assume that $0 \notin \mathfrak{b}$.

A priori estimate **(205)** is given for disturbance f_n in **(440)**, i.e.

$$f_n \in \mathfrak{f} \quad \forall n \geqslant 0 . \qquad (445)$$

Let us introduce designation

$$\mathfrak{X}_{n+1} = \bigcup_{A \in \mathfrak{A},\ b \in \mathfrak{b},\ f_n \in \mathfrak{f}} (AX_n + Bu_n + Cf_n) . \qquad (446)$$

Then we obtain from **(440)** and **(446)** that the motion of system **(440)** is described by the difference equation

$$X_{n+1} \in \mathfrak{X}_{n+1} = \mathfrak{X}(X_n, U_n) . \qquad (447)$$

Let us use hereinafter the characteristic of set $\mathfrak{X}$ introduced above: its distance from the origin of coordinates $\rho_x = \rho(\mathfrak{X})$ and let us consider the objective of control to be the minimization of this quantity by selecting control at each n-th step, i.e. we shall seek control u_n from the solution of the problem

$$\min_{U_n} \{ \rho [\mathfrak{X}_{n+1} = \mathfrak{X}(X_n, u_n)] \} . \qquad (448)$$

From what has been said above it follows that control optimal for the whole class of systems **(440)**, **(441)**, **(445)** is being determined essentially from the solution of the minimax problem.

Taking into account **(440)** and **(446)**, let us rewrite **(447)** in the form

$$\mathfrak{X}_{n+1} = \mathfrak{X}(X_n, u_n) = \sum_{i=1}^{3} \mathfrak{X}_{n+1}^{i} , \qquad (449)$$

where

$$\mathfrak{X}^1_{n+1} = \mathfrak{X}^1(X_n) = \bigcup_{A_m \in \Omega} A(A_m)X_n , \tag{450}$$

$$\mathfrak{X}^2_{n+1} = \mathfrak{X}^2(u_n) = \bigcup_{b \in \mathfrak{b}} B(b)u_n , \tag{451}$$

$$\mathfrak{X}^3_{n+1} = \bigcup_{f_n \in \mathfrak{f}} Cf_n . \tag{452}$$

It will be recalled that in **(449)** and everywhere below, a sum of sets is understood as the Minkowski's sum.

Taking into account the structure of matrix A (ref. to **(203)**), we shall describe set $\mathfrak{X}^1_{n+1}$ in the form

$$\mathfrak{X}^1_{n+1} = \{ X \mid x_i = A^T_i X_n, \quad i = 1,\overline{m-1} , \quad \underline{\pi}(X) \leqslant x_m \leqslant \overline{\pi}(X) \} , \tag{453}$$

where A^T_i is the i-th row of the matrix,

$$\underline{\pi}(X) = \inf_{A_m \in \Omega} \{ A^T_m X_n \} , \tag{454}$$

$$\overline{\pi}(X) = \sup_{A_m \in \Omega} \{ A^T_m X_n \} . \tag{455}$$

In the case if set Ω is a convex polyhedron given by its vertices A^i then as it was already noted above

$$\underline{\pi}(X) = \inf_{i=\overline{1,N}} \{ X^T_n A^i \} , \tag{456}$$

$$\overline{\pi}(X) = \sup_{i=\overline{1,N}} \{ X^T_n A^i \} , \tag{457}$$

where N is the number of vertices of the polyhedron. Having introduced designations

$$\overset{\circ}{\pi}(X) = 0.5(\ \overline{\pi}(X) + \underline{\pi}(X)\ ), \tag{458}$$

$$\Delta_x = 0.5(\ \overline{\pi}(X) - \underline{\pi}(X)\ ), \tag{459}$$

we rewrite (453) in the centered form

$$\mathfrak{X}^1_{n+1}=\{\ X \mid x_i = A_i^T X_n,\quad i = \overline{1,m-1},\quad \overset{\circ}{\pi}(X)-\Delta_x \leqslant x_m \leqslant \overset{\circ}{\pi}(X)+\Delta_x\ \}. \tag{460}$$

Taking into account the structure of vector B, let us write (451) in the form

$$\mathfrak{X}^2_{n+1}=\{\ X \mid x_i = 0,\quad i = \overline{1,m-1},\quad \overset{\circ}{\beta}(u)-\Delta_b u_n \leqslant x_m \leqslant \overset{\circ}{\beta}(u)+\Delta_b u_n\ \}, \tag{461}$$

where

$$\overset{\circ}{\beta}(u) = \overset{\circ}{b}u_n\ . \tag{462}$$

Here

$$\overset{\circ}{b} = 0.5(\ \overline{b} + \underline{b}\ ),\qquad \Delta_b = 0.5(\ \overline{b} - \underline{b}\ ), \tag{463}$$

$$\underline{b} = \min_{b\in\mathfrak{b}} \{b\},\qquad \overline{b} = \max_{b\in\mathfrak{b}} \{b\}. \tag{464}$$

Having described set $\mathfrak{f}$ in the form

$$\mathfrak{f} = \{\ f \mid \overset{\circ}{f} - \Delta_f \leqslant f \leqslant \overset{\circ}{f} + \Delta_f\ \}, \tag{465}$$

and taking into account the structure of vector C, let us rewrite (451) and (449) in the form

$$\mathfrak{X}^3_{n+1}= \{\ X \mid x_i = 0,\quad i = \overline{1,m-1},\quad \overset{\circ}{f} - \Delta_f \leqslant x_m \leqslant \overset{\circ}{f} + \Delta_f\ \}. \tag{466}$$

$$\mathfrak{X}^2_{n+1} = \{\ X \mid x_i = A_i^T X_n\ ,\quad i=\overline{1,m-1}\ ,$$

$$\sigma(X,u) - \delta(X,u) \leqslant x_m \leqslant \sigma(X,u) + \delta(X,u)\ \}, \tag{467}$$

where

$$\sigma(\cdot) = \overset{\circ}{\pi}(X) + \overset{\circ}{b}u_n + \overset{\circ}{f} , \tag{468}$$

$$\delta(\cdot) = \Delta_x + \Delta_f + \Delta_b | u_n | . \tag{469}$$

It follows from (449), (460), (466) and (467) that set $\mathfrak{X}_{n+1}$ is a segment of the straight line parallel to axis Ox_m whose extreme points (ref. to **Fig. 24**) are respectively equal to

$$(X^{\pm}_{n+1})^T = [(\underline{A}X_n)^T, \sigma(X,u) \pm \delta(X,u)] , \tag{470}$$

where $\underline{A}$ is the matrix formed from matrix A by crossing out its last row.

In accordance with the definition of the distance of set $\mathfrak{X}_{n+1}$ from the origin of coordinates introduced above, its minimum is attained on condition that $\| X^-_{n+1} \| = \| X^+_{n+1} \|$ which in its turn takes place only when the condition is fulfilled

$$\sigma(X,u) = \overset{\circ}{\pi}(X) + \overset{\circ}{b}u_n + \overset{\circ}{f} = 0 . \tag{471}$$

Substituting here the value of $\overset{\circ}{\pi}(X)$ from (457) and (455), (456) and taking into account designation (463), we obtain

$$0.5\{ \inf_{A_m \in \Omega} (A^T_m X_n) + \sup_{A_m \in \Omega} (A^T_m X_n) + (\underline{b} + \bar{b})u_n \} + \overset{\circ}{f} = 0 . \tag{472}$$

From where follows directly

<u>Statement 7.</u> Control $\overset{*}{u}_n$ equal to

$$u_n = -\frac{0.5[\inf (A^T_m X_n) + \sup (A^T_m X_n)] + \overset{\circ}{f}}{0.5(\underline{b} + \bar{b})} \tag{473}$$

is the solution of problem (448) and therefore it is optimal for the whole class of systems (440), (441).

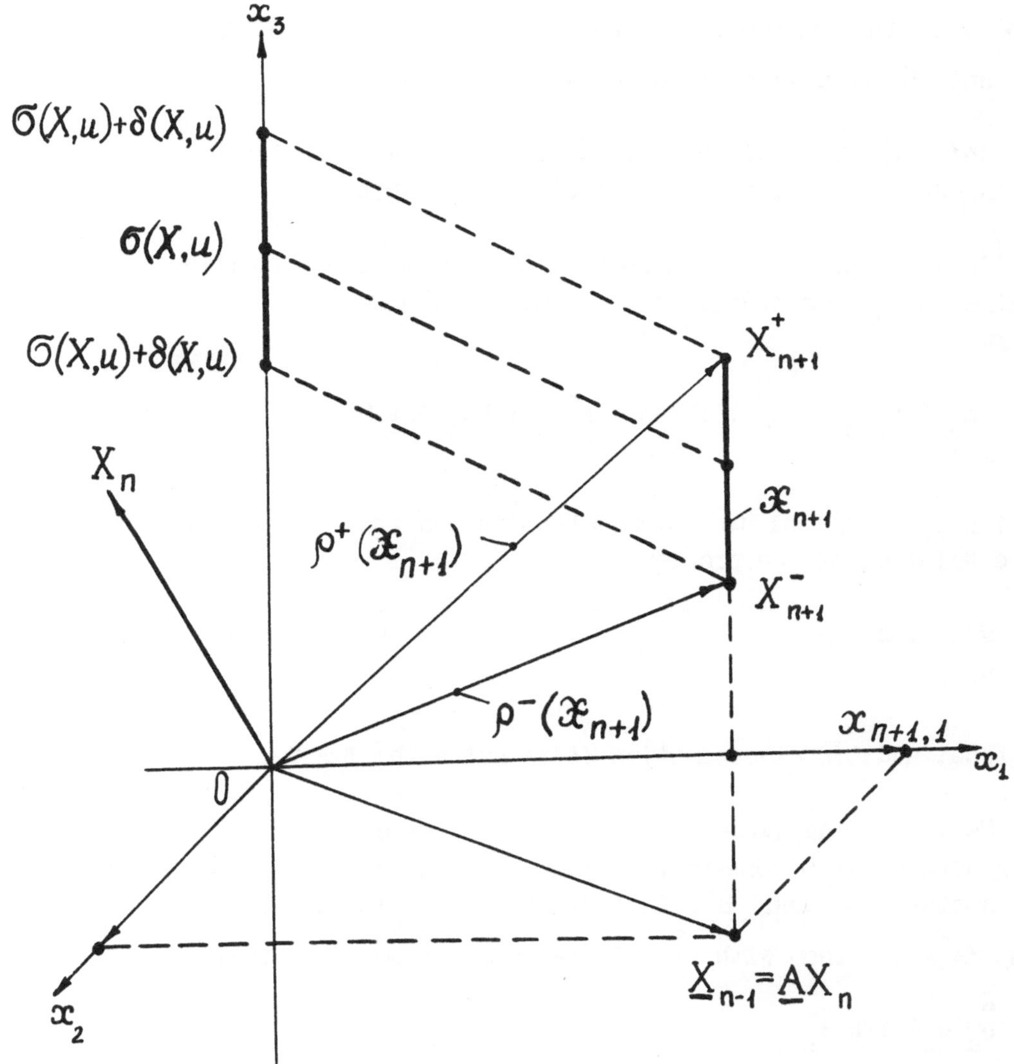

Fig. 24. Construction of set $\mathfrak{X}_{n+1}$

Let us consider the solution of the synthesis problem formulated above in that special case when $\overset{\circ}{f} = 0$. Substituting the values of $\underline{b}$ and $\bar{b}$ from (464) into (473), we obtain

$$\inf_{A_m \in \mathfrak{A}} (A_m^T X_n) + \inf_{b \in \mathfrak{b}} (bu_n) + \sup_{A_m \in \mathfrak{A}} (A_m^T X_n) + \sup_{b \in \mathfrak{b}} (bu_n) = 0 \ . \tag{474}$$

In view of the independence (according to definition (441)) of estimates of the values of A_m and b , let us rewrite (474) in the form

$$\inf_{L \in \mathfrak{L}} \{ A_m^T X_n + bu_n \} + \sup_{L \in \mathfrak{L}} \{ A_m^T X_n + bu_n \} = 0 \ ,$$

which is identical to the result obtained above (ref. to (225)) from the solution of the problem

$$\min_{u_n} \max_{L \in \mathfrak{L}} \{ \omega_{n+1} = | x_{m,n+1} | = | A_m^T X_n + bu_n + f_n | \}$$

without additional assumption (441) but with $\overset{\circ}{f} = 0$.

Next, let us consider the limiting case when sets $\mathfrak{A}$ and $\mathfrak{b}$ degenerate into one-point sets comprising only true values of parameters A and b . In this case $\underline{b} = \bar{b} = b$ and $\inf (A_m^T X_n) = \sup (A_m^T X_n)$. Then with $\overset{\circ}{f} = 0$ we obtain from (472) that

$$\overset{*}{u}_n = - 1/b \ A_m^T X_n \ . \tag{475}$$

Simultaneous solution of (475) and (440) gives

$$X_{n+1} = \tilde{A} X_n + Cf_n \ , \qquad \text{where } \tilde{A} = \left\| \begin{array}{c|c} 0 & I_{m-1} \\ \hline \multicolumn{2}{c}{0} \end{array} \right\| . \tag{476}$$

Thus, optimal control (473) in the case being considered generates the matrix of the closed system in the form of the nilpotent matrix. When this takes place for an autonomous system i.e. with $f = 0$, this control provides for system **(440)** not only its asymptotic stability but also the transient process of finite duration.

Let us consider now the possibility to reduce the volume of computations required to determine control u_n^* in accordance with **(473)**. The most laborious operations in **(473)** are the operations of determining **inf** and **sup** of scalar products $A_m^T X_n$. For controlled plants with comparatively small dimensionality (m is about 5 to 7), the simplest way to determine **inf** and **sup** for sets Ω_m in the form of polyhedra is to use the method of exhaustive search. But for the systems of higher order when the number of vertices of polyhedron measures already tens and hundreds, the solution requires already the application of more efficient methods and nevertheless it will be also require sufficiently big computing efforts. A radical reduction of the volume of computations can be achieved only with the change to the linear suboptimal control of the form **(475)** where instead of vector A_m any other vector $\tilde{A}_m \in \Omega_m$ can be taken.

APPENDIX A

REDUCTION OF DIFFERENCE EQUATIONS OF GENERAL FORM TO CANONICAL FORM

Let a linear continuous dynamic plant be given with scalar control $u(t)$ representing a sequence of amplitude-modulated standard pulses with amplitude u_n and constant pulse repetition period $T = const$, i.e.

$$u(t) = u_n \qquad \forall\, t \in [\, nT,\ (n+1)T] \,. \tag{A.1}$$

Since such control $u(t)$ is a discontinuous function of time then the description of the motion of the linear dynamic plant in terms of differential equations is unconvenient and because of this in order to simplify its motion it is appropriate to use difference equations describing plant motion only at discrete instants of time. Difference equation for a linear continuous dynamic plant of the m-th order subjected to the action of a scalar disturbance has the form

$$A(E)y_n = B(E)u_n + f_{n+m-1} \,, \qquad n = 0,\ 1,\ 2,\ \dots \,, \tag{A.2}$$

where y_n is the value of the output scalar coordinate being measured, u_n is control, f_{n+m-1} is the external disturbance where

$$A(E) = \sum_{i=0}^{m} a_i E^{m-i} \,, \qquad B(E) = \sum_{j=1}^{s} b_j E^{s-j} \,,$$

a_i and b_j are known numeric coefficients, $s \leqslant m$. Shift (to the right) operator E was used in writing equation (A.2) which is defined as $Ey_n = y_{n+1}$. Hereinafter it is assumed without the loss of generality that $a_0 = 1$. Besides, let us assume that $b_s \neq 0$.

Let us note that it is this form in which difference equations of continuous dynamic plants are obtained using standard techniques.

Hereinafter it is convenient to change from difference equations written in form **(A.2)** to the other form of their writing. Multiplying the right-hand and left-hand sides of equation **(A.2)** by $E^{-(m-1)}$ and introducing a new shift (to the left) operator D defined as $Dy_n = y_{n-1}$, we obtain

$$A(D)y_n = B(D)u_n + f_n , \qquad n = 0, 1, \dots , \tag{A.3}$$

where

$$A(E) = \sum_{i=0}^{m} a_i D^{i-1} , \qquad B(E) = \sum_{j=1}^{s} b_j D^{j-1} .$$

Initial conditions $y_0, y_{-1}, \dots, y_{-(m-1)}$ and $u_{-1}, u_{-2}, \dots, u_{-(s-1)}$ are given for equation **(A.3)** at $n = 0$. In the subsequent discussion assume that measured at the n-th instant of time is the value of the output coordinate y_n which along with the values $y_{n-1}, y_{n-2}, \dots, y_{n-(m-1)}$ measured earlier and stored in memory registers as well as with the values of control realized earlier $u_{n-1}, u_{n-2}, \dots, u_{n-s+1}$ are used to generate control u_n , i.e. in the general case $u_n = u(y_n, y_{n-1}, \dots, y_{n+m-1}, u_{n-1}, u_{n-2}, \dots, u_{n-s+1})$.

First let us consider the case when $s = 1$ and $b_1 \neq 0$. Then introducing new variables

$$x_{i,n} = y_{n+m-1} \qquad \forall\ i \in [\,1, m] , \tag{A.4}$$

let us write equation **(A.3)** in vector-matrix form

$$X_{n+1} = AX_n + Bu_n + \Gamma f_n , \quad X_0 = \overset{\circ}{X} , \qquad n = 0, 1, \dots , \tag{A.5}$$

where

$$X_n = \begin{Vmatrix} x_{1,n} \\ x_{2,n} \\ \vdots \\ x_{m-1,n} \\ x_{m,n} \end{Vmatrix}, \quad B = \begin{Vmatrix} 0 \\ 0 \\ \vdots \\ 0 \\ 1 \end{Vmatrix}, \quad \Gamma = \begin{Vmatrix} 0 \\ 0 \\ \vdots \\ 0 \\ 1 \end{Vmatrix}, \quad A = \begin{Vmatrix} 0 & I_{m-1} \\ \hline -a_m \; -a_{m-1} & \ldots \; -a_1 \end{Vmatrix}.$$

I_{m-1} is unitary matrix $(m-1) \times (m-1)$. Since condition $b_1 \neq o$ is the controllability condition for system (A.5) then it obvious that at $f \equiv 0$ there exists control $u_n = u(X_n)$ which stabilizes system (A.5).

Let us note that as follows from equations (A.4), state vector X_n consists only of those components which are already measured and known at the n-th discrete instant of time. Because of this, the problem of so-called structural restrictions does not exists for system (A.5) at $n \geqslant m$.

Let us consider now a more general case when $s > 1$ and $b_1 \neq 0$ as before. In this case, along with variables $x_{1,n}, x_{2,n}, \ldots, x_{m,n}$ (ref. to (A.4)), let us introduce variables

$$x_{m+j,n} = u_{n-s+j} \qquad \forall\, j \in [\,1,\, s-1] \,. \tag{A.6}$$

Then we obtain system of difference equations

$$x_{i,n+1} = x_{i+1,n} \qquad \forall\, i \in [\,1,\, m-1] \,,$$

$$x_{m,n+1} = -a_m x_{1,n} - a_{m-1} x_{2,n} - \ldots - a_2 x_{m-1,n} - a_1 x_{m,n} +$$

$$+ b_s x_{m+1,n} + b_{s-1} x_{m+2,n} + \ldots + b_2 x_{m+s-1} + b_1 u_n + f_n \,, \tag{A.7}$$

$$x_{i,n+1} = x_{i+1,n} \qquad \forall\, i \in [\,m+1,\, m+s-2] \,,$$

$$x_{m+s-1,n+1} = u_n \,,$$

which after introduction of

$$\tilde{X}_n = \left\| \begin{matrix} x_{1,n} \\ x_{2,n} \\ \cdot \\ \cdot \\ \cdot \\ x_{m+1,n} \\ x_{m+1,n} \\ \cdot \\ \cdot \\ \cdot \\ x_{m+s-1,n} \end{matrix} \right\| \begin{matrix} \left.\vphantom{\begin{matrix} x \\ x \\ \cdot \\ \cdot \\ \cdot \\ x \end{matrix}}\right\} m \\ \left.\vphantom{\begin{matrix} x \\ \cdot \\ \cdot \\ \cdot \\ x \end{matrix}}\right\} s-1 \end{matrix}, \quad \tilde{B} = \left\| \begin{matrix} 0 \\ 0 \\ \cdot \\ \cdot \\ \cdot \\ 0 \\ b_1 \\ 0 \\ 0 \\ \cdot \\ \cdot \\ \cdot \\ 0 \\ 1 \end{matrix} \right\| \begin{matrix} \} m \\ \\ \} s-1 \end{matrix}, \quad \tilde{\Gamma} = \left\| \begin{matrix} 0 \\ 0 \\ \cdot \\ \cdot \\ \cdot \\ 0 \\ 1 \\ 0 \\ 0 \\ \cdot \\ \cdot \\ \cdot \\ 0 \end{matrix} \right\| \begin{matrix} \} m \\ \\ \} s-1 \end{matrix},$$

$$\tilde{A} = \left\| \begin{array}{c|c||c|c} \overbrace{\begin{matrix} 0 \\ \cdot \\ \cdot \\ \cdot \\ 0 \end{matrix} \quad\quad I_{m-1}}^{m} & & \overbrace{\quad 0 \quad}^{s-1} & \\ \hline -a_m & -a_{m-1} \ldots -a_1 & b_s & b_{s-1} \ldots b_2 \\ \hline\hline & & \begin{matrix} 0 \\ \cdot \\ \cdot \\ \cdot \\ 0 \end{matrix} & \begin{matrix} I_{s-2} \\ \\ 0 \ldots 0 \end{matrix} \\ \multicolumn{2}{c||}{0} & & \end{array} \right\| \begin{matrix} \} m \\ \\ \} s-1 \end{matrix},$$

can be written in the form of vector-matrix equation

$$\tilde{X}_{n+1} = \tilde{A}\tilde{X}_n + \tilde{B}u_n + \tilde{\Gamma}f_n \,, \quad \tilde{X}_0 = \tilde{X}^0 \,, \quad n = 0, 1, \ldots . \tag{A.8}$$

Since matrix $\mathbf{A}$ in this equation has no canonical form then equation (A.8) is inconvenient for solving analysis and synthesis problems. Because of this, let us introduce new variables

$$z_{i,n} = C^T\tilde{A}^{i-1}\tilde{X}_n , \qquad i \in [\ 1,\ m+s-1\ ] , \tag{A.9}$$

where $C^T = (\ c_1,\ c_2,\ \dots,\ c_{m-1},\ 1,\ c_{m+1},\ \dots,\ c_{m+s-1})$ and components c_j $(j \neq m)$ of vector C are determined from the solution of the bilinear equation system

$$C^T\tilde{A}^{j-1}\tilde{B} = 0 \qquad \forall\ j \in [\ 1,\ m+s-2\ ] . \tag{A.10}$$

Let us rewrite equations **(A.9)** in vector-matrix form

$$Z_n = H\tilde{X}_n , \tag{A.11}$$

where $Z_n^T = (\ z_{1,n},\ z_{2,n},\ \dots,\ z_{m+s-1,n}\ )$ and

$$H = \left\| \begin{matrix} C \\ C^T\tilde{A} \\ \cdot \\ \cdot \\ \cdot \\ C^T\tilde{A}^{m+s-2} \end{matrix} \right\| . \tag{A.12}$$

Let us suppose that matrix H is non-singular (i.e. that system **(A.8)** is controllable). Then equation **(A.8)** can be rewritten in the form

$$Z_{n+1} = AZ_n + Bu_n + \Gamma f_n , \quad Z_0 = \overset{\circ}{Z} , \quad n = 0,\ 1,\ \dots , \tag{A.13}$$

where

$$Z_n = \left\| \begin{matrix} z_{1,n} \\ z_{2,n} \\ \cdot \\ \cdot \\ \cdot \\ z_{m+s-1,n} \end{matrix} \right\| , \qquad \check{B} = \left\| \begin{matrix} 0 \\ 0 \\ \cdot \\ \cdot \\ \cdot \\ 0 \\ \check{b} \end{matrix} \right\| , \qquad \check{\Gamma} = \left\| \begin{matrix} \check{\gamma}_1 \\ \check{\gamma}_2 \\ \cdot \\ \cdot \\ \cdot \\ \check{\gamma}_{m+s-1} \end{matrix} \right\| ,$$

$$\check{b} = C^T\tilde{A}^{m+s-2}\tilde{B}\ , \qquad \tilde{\gamma}_i = C^T\tilde{A}^{i-1}\tilde{\Gamma} \qquad \forall\ i \in [\ 1,\ m+s-1],\qquad (A.14)$$

$$\check{A} = \left\| \begin{array}{c|c} 0 & \\ \cdot & \\ \cdot & I_{m+s-2} \\ \cdot & \\ 0 & \\ \hline \check{a}_1 & \check{a}_2\ \ldots\ \check{a}_{m+s-1} \end{array} \right\| ,$$

and $\check{a}_i$ is the i-th component of vector $(H^T)^{-1}(\tilde{A}^T)^{m+s-1}C$.

Let us note that system of equations **(A.10)** is not a system of homogeneous equations. The non-homogenity of the system follows from the fact that as a result of multiplication of the number $c_m = 1$ by the m-th component of vector $\tilde{A}^{i-1}\tilde{B}$ we obtain generally (with controllable pair $\tilde{A}$ and $\tilde{B}$) a non-zero number which is the free term of the i-th equation of system **(A.10)**. Then in obtaining equation **(A.13)** we have taken into account that

$$z_{i,n+1} = z_{i+1,n} + C^T\tilde{A}^{i-1}\tilde{\Gamma}f_n \qquad \forall\ i \in [\ 1,\ m+s-2]\ .$$

It follows from **(A.14)** that at $\check{b} \neq 0$ (this inequality is the condition of controllability of system **(A.13)** and, therefore, of system **(A.8)**) $f_n \equiv 0$ and there exists control u_n of the form $u_n = u(Z_n)$ which stabilizes system **(A.13)**. Hence, there exists also control $u_n = u(H\tilde{X}_n)$ which stabilizes system **(A.8)** at $f_n \equiv 0$.

Let us emphasize here that since this conclusion was obtained without introducing any restrictions on the numerical values of coefficients b_j , $j \in [\ 2,\ s]$ and, therefore, on the values of the roots of equation

$$B(E) = 0\ , \qquad (A.15)$$

then the conclusion on the stabilizability of system **(A.8)** holds true also for non-minimum-phase controlled plants **(A.2)**, i.e. for such plants for which there exists at least one root E_i of equation (A.15) which does not satisfy condition $|E_i| < 1$.

Let us consider now a case commonly encountered in practice when due to one or other reason (in particular, due to the presence of a pure delay in the controlled plant or in the controlling unit) values $y_n, y_{n-1}, \ldots, y_{n-k}$ ($k < m-1$) can not be used in generating control. In this case control of the form $u_n = u(H\tilde{X}_n)$ is physically not feasible since a part of components of vector $\tilde{X}_n$, namely $x_{m,n}$, $x_{m-1,n}$, ..., $x_{m-k,n}$ is not measured. To remove structural restrictions of such kind, let us use the following technique. Due to equation (A.3), let us express non-measured coordinates $x_{m-j,n} = y_{n-j}$ ($j \in [1, k]$) in terms of values of $y_{n-(k+1)}, y_{n-(k+2)}, \ldots, y_{n-(k+m-1)}$ stored in memory register as well as the values of control which are already realized $u_{n-1}, u_{n-2}, \ldots, u_{n-(k+s-1)}$

$$x_{m-j,n} = D^{j+1}[\tilde{A}(D)y_n + B(D)u_n + f_n] , \qquad j \in [1,k] , \tag{A.16}$$

where

$$\tilde{A}(D) = -\sum_{i=1}^{m} a_i D^{-(i-1)} .$$

Using for generation of control of the form $u_n = u(H\tilde{X}_n)$ instead of non-measured coordinates $x_{m-j,n}$ their estimates obtained from (A.16) at $D^{j+1}f_n \equiv 0$ for all $j \in [1,k]$, we obtain control which stabilizes system (A.8) in the absence of disturbances, i.e. at $f_n \equiv 0$ even in the case when the mentioned part of coordinates is not measured, i.e. in the presence of structural restrictions.

REFERENCES

1. Eykhoff, P. System Identification Parameter and State Estimation.. John Willey and Sons, New York, 1974.

2. Zypkin, Ya. Z., Adaptation and Learning in Automatic systems. Nauka, Moscow, 1968 (in Russian).

3. Aström, K. J., Eykhoff, P. System Identification. A Survey. Automatica, No. 7, 1971.

4. Mendel, J. M. Discrete Techniques of Parameter Estimation. The Equation Error Formulation. Marcel Dekker, N.Y., 1973.

5. Graupe, D. Identification of Systems. Robert E. Krieger Publishing Company, Huntington, N.Y., 1978.

6. Ljung, L., Söderström, T. Theory and Practice of Recursive Identification. MIT Press, Cambridge, Mass., 1983.

7. Tsypkin, Ya. Z. Foundations of Informational Theory of Identification. Nauka, Moscow, 1984 (in Russian).

8. Kuntzevich, V. M., Lychak, M. M. On Optimal and Adaptive Control of Dynamic Plants under Conditions of Uncertainty. Automation and Remote Control, No. 1, 1979 (in Russian).

9. Kuntzevich, V. M., Lychak M. M. Obtaining Guaranteed Estimates in Parametric Identification Problems. Soviet Automatic Control, No. 4, 1982.

10. Kuntzevich, V. M., Lychak, M. M. Synthesis of Optimal and Adaptive Control Systems. Game Approach. Naukova Dumka, Kiev, 1985 (in Russian).

11. Kuntzevich, V. M., Lychak, M. M., Nikitenko, A. S. Obtaining Estimates in Form of Sets in Parametric Identification Problem. 8th IFAC/IFORS Symposium, Beijing, 1988.

12. Schwepe, F. G. Uncertain Dynamic systems. Prentice Hall, Englewood, Cliffs., New Jersey, 1973.

13. Chernousko, F. L. Optimal Guaranteed Estimates of Uncertainties by means of Ellipsoids. Izvestiya AN SSSR. Technicheskaya Kibernetika, 1980, No. 4, No. 5, (in Russian).

14. Chernousko, F. L. Estimation of Phase State of Dynamic Systems. Nauka, Moscow, 1988 (in Russian).

15. Kurzhanski, A. B. Identification - A Theory of Guaranteed Estimates. IIASA Working Paper. Laxenburg, Austria, 1989.

16. Bakan, G. M. Filtering under Conditions of Non-Statistically Set Uncertainty. Soviet Automatic Control, No. 2, 1980 (in Russian).

17. Bakan, G. M. Non-Statistical Statement and Solution of one Filtering Problem. Automation and Remote Control, No. 9, 1983 (in Russian).

18. Kuntzevich, V. M., Lychak, M. M., Nikitenko, A. S. Solution of a System of Linear Equations Given Uncertainty in its Both Sides. Cybernetics, No. 4, 1988 (in Russian).

19. Kalman, R. E. Identification of Noisy Systems. Uspekhi Mathematitsheskikh Nauk, Vol. 10, No. 4, 1984 (in Russian).

20. Moore, R. E. Interval analysis. Prentice Hall, Englewood, Cliffs, New Jersey, 1966.

21. Interval mathematics, ed. by K. Nickel, Lecture notes in computer science, 29. Springer Verlag, Berlin-Heidelberg, 1975.

22. Kalmikov, S. A., Shokin, Yu. J., Yuldashev, Z. Kh. Methods of Interval Analysis. Nauka, Novosibirsk, 1986 (in Russian).

23. Kuntzevich, V. M., Lychak, M. M. Elements of Theory of Set Evolution and Stability of These Processes. Cybernetics, No. 1, 1983, (in Russian).

24. Beeck, H. Über Structur und Abschätzungen der Lösungsmenge von linearen Gleichungssystem mit Intervallkoeffizienten. Computing, 10, 1972.

25. Kuntzevich, V. M., Lychak, M. M., Nikitenko, A. S. Active Identification of Static Plant Parameters (Game Problem of Experimental Design). Automation and Remote Control, No.9, 1987 (in Russian).

26. Kuntzevich, V. M. Optimal Control of Discrete Dynamic Plants with Unknown Nonstationary Parameters. Automation and Remote Control, No. 2, 1980 (in Russian).

27. Kuntzevich, V. M. Determiniмg Guaranteed Estimates of State and Parameter Vectors with Restricted Disturbances. Doklady AN SSSR, Vol. 288, No. 3, 1986 (in Russian).

28. Kuntzevich, V. M. On Simultaneous Construction of Guaranteed Estimates of State and Parameter Vectors of Discrete Control Systems with Restricted Disturbances and Noise. Cybernetics and Computer Engineering. Naukova Dumka, Kiev, Vol. 87, 1990 (in Russian).

29. Gay, D. M. Solving Interval Linear Equations. SIAM J. Numer. Anal., Vol. 19, No. 4, 1982.

30. Lasdon, L. S. Optimization Theory for Large Systems. MacMillan Company, London, 1978.

31. Tsypkin, Ya. Z. Foundations of Learning Systems Theory. Nauka, Moscow, 1970 (in Russian).

32. Aoki, T. Stochastic Systems Optimization. Nauka, Moscow, 1971 (in Russian).

33. Petrov, B. N., Rutkovski, I. Yu., Krutova, I. N., Zemlakov, S. D. Principles of Construction and Design of Self-Adjusting Control Systems. Moscow, Mashinostroyenie, 1972 (in Russian).

34. Saridis, G. N. Self-Organizing Control of Stochastic Systems, Marcel Dekker, Inc. New York and Basel, 1977.

35. Derevitski, D. P., Fradkov, A. L. Applied Theory of Discrete Adaptive Control Systems. Nauka, Moscow, 1981 (in Russian).

36. Rastrigin, L. A. Adaptation of Complex Systems. Zinate, Riga, 1981 (in Russian).

37. Sragovich, V. G. Adaptive Control. Nauka, Moscow, 1981 (in Russian).

38. Fomin, V. N., Fradkov, A. L., Yakubovich, V. A. Adaptive Control of Dynamic Plants. Nauka, Moscow, 1981 (in Russian).

39. Landau, J. D. Adaptive Control: the Model Reference Approach. Control and Systems Theory Series, 1979, Vol. 8, Marcel Dekker, Inc.

40. Moiseev N. N., Ivanilov, Yu. P., Stolarov, E. M. Optimization Methods. Naukova Dumka, Kiev, 1978 (in Russian).

41. Mikhalevich, V. S., Volkovich, V. L. Computig Methods of Investigation and Design of Complex Systems. Nauka, Moscow,1982 (in Russian).

42. Emelyanov, S. V., Larichev, O. I. Multiple Criteria Methods of Decision Making. Ekonomika, Moscow, 1984 (in Russian).

43. Chankeng, V., Haimes, J. J. Multiobjective Decision Making: Theory and Methodology, New York, North-Holland, 1983.

44. Steuer, R. Multiple Criteria Optimization: Theory, Computation and Application, John Wiley and Sons, New York, 1986.

45. Yudin, D. B. Computing Methods of Decision Making Theory. Nauka, Moscow, 1986 (in Russian).

46. Krasovski, N. N. Gaming Problems of Motion Encounter. Nauka, Moscow, 1970 (in Russian).

47. Krasovski, N. N., Subbotin, A. I. Position Differential Games. Nauka, Moscow, 1974 (in Russian).

48. Kuzzhanski, A. B. Control and Observations under Uncertainty Conditions. Nauka, Moscow, 1981 (in Russian).

49. Chernousko, F. L., Melikyan, A. A. Game Problems of Control and Search. Nauka, Moscow, 1978 (in Russian).

50. Keyn, V. M. Optimization of Control Systems by Minimax Criterion. Nauka, Moscow, 1981 (in Russian).

51. Demyanov, V. F., Malozemov, V. N. Introduction to Minimax. Nauka, Moscow, 1972 (in Russian).

52. Theory Problems and Elements of Software of Minimax Problems. Ed. by V.F. Demyanov, V.N. Malozemov. Leningrad University Press, Leningrad, 1977 (in Russian).

53. Fedorov, V. V. Numerical Methods of Maximin. Nauka, Moscow, 1979 (in Russian).

54. Ortega, J. M., Reinboldt, W. C. Iterative Solution of Nonlinear Equations in Several Variables. Academic Press, New York, 1970

55. Hamming, R. W. Numerical Methods for Scientists and Engineers. 2 ed., Mc-Graw-Hill, New York, 1973.

56. Dennis, J. E., Schnabel, R. B. Numerical Methods for Unconstrained Optimization and Nonlinear Equations, Prentice-Hall, Englewood Cliffs, New Jersey, 1983.

57. Gass, S. I. Linear Programming: Methods and Applications, New Jork: McGraw-Hill, 1985.

58. Zionts, S. Linear Programming. Englewood Cliffs, New Jersey: Prentice-Hall, 1974.

59. Kuntzevich, V. M., Lychak, M. M. Synthesis of Automatic Control Systems by Means of Lyapunov Functions. Nauka, Moscow, 1977 (in Russian).

60. Foundations of Automatic Control. Theory. Ed. by V.V. Solodovnikov. Mashinostroyenie, Moscow, 1954 (in Russian).

61. Oppelt, W. Kleines Handbuch Technischer Regelvorgänge. 5, Auflage, Verlag Chemie, Wienheim, 1972.

62. Yakubovich, V. A. Method of Recursive Objective Inequalities in Adaptive Systems Theory. In Problems of Cybernetics, Adaptive Systems. Scientific Council on Cybernetics. USSR Acad. Sci., 1976 (in Russian).

63. Bondarenko, V. A., Yakubovich, V. A. Method of Recursive Objective Inequalities in Adaptive Systems Theory: Results and Problems. In Problems of Cybernetics. Problems and Methods of Adaptive Conterol. Scientific Council on Cybernetics, USSR Acad. Sci., Moscow, 1981 (in Russian).

64. Tsypkin, Ya. Z. Theory of Linear Impulse Systems. Fizmatgiz, Moscow, 1963 (in Russian).

65. Jury, E. Impulse Systems of Automatic Control. Fizmatgiz, Moscow, 1969 (in Russian).

66. Thoma, M. Theorie Linearer Regelsysteme. Vieweg, Braunschweig, 1973.

67. Föllinger, O. Lineare Abtast-Systeme, R. Oldenbourg Verlag, 1973.

68. Aizerman, M. A., Litvakov, B. M. Pseudocriteria and Pseudocriterial Chiose. Mathematical Social Sciences, 1989, vol. 176.

69. Siliak, D. D. Parameter Space Methods for Robust Control Design: a Guided Tour. IEEE Tranc. on Automatic Control, Vol. 34, No. 7, 1989.

70. Kuntzevich, V. M., Platova, E. L. Robust Stability of Continuous and Discrete Systems with Parametrically Given Estimates of Their Coefficients. Automatika, No. 2, 1991 (in Russian).

71. Kharitonov, V. L. On Asymptotic Stability of Equilibrium Position of a Family of Linear Differential Equation Systems. Diferentzialniye Uravneniya, 14, No. 11, 1978 (in Russian).

72. Ackermann, J., Anderson, B. D., Hollot, C. V., Khargonekar, P. P. Discussion. New Trends in Robustness Analysis, 28-th IEE Conf. on Decision and Control, 1989.

73. Kiendl, H., Ossadnik, H. Robuste Ljapunov-Functionen und Anwendung für die Robustheitsanalyse und Synthese Mit der Methode der Konvexen Zerlegung. Dritter Workshop. Anforderungsspecifische Entwurfsmethoden, Interlaken 1989, Kurzfassung in Automatisierungstechnik, 1989.

74. Barmish, B. H. A Generalization of Kharitonov's four Polynomial Concept for Robust Stability Problems with Lineary Depended Coefficient Perturbation, IEEE Trans. Automat. Control, Vol. AS-34, No. 2, 1989.

75. Polak, B. T., Tsypkin, Ya. Z. Frequency Criteria of Robust Stability and Aperiodicity of Linear Systems. Automatika i Telemekhanika, No. 4, 1990 (in Russian).

76. Hollot, C. V., Bartlett, A. C. Some Discrete Couterparts to Kharitonov's Stability Criteria for Uncertain Systems, IEEE Trans. Automat. Control, Vol. AC-31, No. 4, 1986.

77. Ackermann, J. E., Hu, H. Z. Robustness of Sampled-Data Control Systems with Uncertain Physical Plant Parameters, 11-th IFAC World Congress, Tallin, Vol. 5, 1990.

78. Jury, E. I. Robustness of Discrete Systems. A Review. Automatika i Telemekhanika, No. 5, 1990 (in Russian).

79. Shor, N. Z. Minimization Methods for Non-Differentiable Functions, Springer-Verlag, Heidelberg, 1985.

80. Kuntzevich, A. V. To the Problem of the Efficiency of the Application of Optimization Algorithms with Space Extension, Kibernetika, No. 2, 1989 (in Russian).

81. Kuntzevich, A. V., Kuntzevich, V. M. Instrumental System "Robust Stability" for Analysis of the Robust Stability of Dynamic Systems, Automatika, No. 6, 1990 (in Russian).

82. Aubin, J. P., Celenia, A. Differential Inclusions, Springer-Verlag, Heidelberg, 1984.

INDEX

Lecture Notes in Control and Information Sciences

Edited by M. Thoma and A. Wyner

Vol. 117: K.J. Hunt
Stochastic Optimal Control Theory
with Application in Self-Tuning Control
X, 308 pages, 1989.

Vol. 118: L. Dai
Singular Control Systems
IX, 332 pages, 1989

Vol. 119: T. Başar, P. Bernhard
Differential Games and Applications
VII, 201 pages, 1989

Vol. 120: L. Trave, A. Titli, A. M. Tarras
Large Scale Systems:
Decentralization, Structure Constraints
and Fixed Modes
XIV, 384 pages, 1989

Vol. 121: A. Blaquière (Editor)
Modeling and Control of Systems
in Engineering, Quantum Mechanics,
Economics and Biosciences
Proceedings of the Bellman Continuum
Workshop 1988, June 13–14, Sophia Antipolis, France
XXVI, 519 pages, 1989

Vol. 122: J. Descusse, M. Fliess, A. Isidori,
D. Leborgne (Eds.)
New Trends in Nonlinear Control Theory
Proceedings of an International
Conference on Nonlinear Systems,
Nantes, France, June 13–17, 1988
VIII, 528 pages, 1989

Vol. 123: C. W. de Silva, A. G. J. MacFarlane
Knowledge-Based Control with
Application to Robots
X, 196 pages, 1989

Vol. 124: A. A. Bahnasawi, M. S. Mahmoud
Control of Partially-Known
Dynamical Systems
XI, 228 pages, 1989

Vol. 125: J. Simon (Ed.)
Control of Boundaries and Stabilization
Proceedings of the IFIP WG 7.2 Conference
Clermont Ferrand, France, June 20–23, 1988
IX, 266 pages, 1989

Vol. 126: N. Christopeit, K. Helmes
M. Kohlmann (Eds.)
Stochastic Differential Systems
Proceedings of the 4th Bad Honnef Conference
June 20–24, 1988
IX, 342 pages, 1989

Vol.127: C. Heij
Deterministic Identification
of Dynamical Systems
VI, 292 pages, 1989

Vol. 128: G. Einarsson, T. Ericson,
I. Ingemarsson, R. Johannesson,
K. Zigangirov, C.-E. Sundberg
Topics in Coding Theory
VII, 176 pages, 1989

Vol. 129: W. A.Porter, S. C. Kak (Eds.)
Advances in Communications and
Signal Processing
VI, 376 pages, 1989.

Vol. 130: W. A. Porter, S. C. Kak,
J. L. Aravena (Eds.)
Advances in Computing and Control
VI, 367 pages, 1989

Vol. 131: S. M. Joshi
Control of Large Flexible Space Structures
IX, 196 pages, 1989.

Vol. 132: W.-Y. Ng
Interactive Multi-Objective Programming
as a Framework for Computer-Aided Control
System Design
XV, 182 pages, 1989.

Vol. 133: R. P. Leland
Stochastic Models for Laser Propagation
in Atmospheric Turbulence
VII, 145 pages, 1989.

Vol. 134: X. J. Zhang
Auxiliary Signal Design in Fault
Detection and Diagnosis
XII, 213 pages, 1989

Vol. 135: H. Nijmeijer, J. M. Schumacher (Eds.)
Three Decades of Mathematical System Theory
A Collection of Surveys at the Occasion of the
50th Birthday of Jan C. Willems
VI, 562 pages, 1989

Vol. 136: J. Zabczyk (Ed.)
Stochastic Systems and Optimization
Proceedings of the 6th IFIP WG 7.1
Working Conference,
Warsaw, Poland, September 12–16, 1988
VI, 374 pages. 1989

Lecture Notes in Control and Information Sciences

Edited by M. Thoma and A. Wyner

Vol. 137: S. L. Shah, G. Dumont (Eds.)
Adaptive Control Strategies for Industrial Use
Proceedings of a Workshop
Kananaskis, Canada, 1988
VI, 360 pages. 1989

Vol. 138: D. C. McFarlane, K. Glover
Robust Controller Design Using Normalized Coprime Factor Plant Descriptions
X, 206 pages. 1990

Vol. 139: V. Hayward, O. Khatib (Eds.)
Experimental Robotics I
The First International Symposium
Montreal, June 19-21, 1989
XVII, 613 pages. 1990

Vol. 140: Z. Gajic, D. Petkovski, X. Shen
Singularly Perturbed and Weakly Coupled Linear Control Systems
A Recursive Approach
VII, 202 pages. 1990

Vol. 141: S. Gutman
Root Clustering in Parameter Space
VIII, 153 pages. 1990

Vol. 142: A. N. Gündeş, C. A. Desoer
Algebraic Theory of Linear Feedback Systems with Full and Decentralized Compensators
V, 176 pages. 1990

Vol. 143: H.-J. Sebastian, K. Tammer (Eds.)
System Modelling and Optimization
Proceedings of the 14th IFIP Conference
Leipzig, GDR, July 3-7, 1989
X, 960 pages. 1990

Vol. 144: A. Bensoussan, J. L. Lions (Eds.)
Analysis and Optimization of Systems
Proceedings of the 9th International Conference, Antibes, June 12-15, 1990
XII, 992 pages. 1990

Vol. 145: M. B. Subrahmanyam
Optimal Control with a Worst-Case Performance Criterion and Applications
X, 133 pages. 1990

Vol. 146: D. Mustafa, K. Glover
Minimum Entropy H_∞ Control
IX, 144 pages. 1990

Vol. 147: J. P. Zolésio (Ed.)
Stabilization of Flexible Structures
Third Working Conference, Montpellier, France, January 1989
V, 327 pages, 1990

Vol. 148: In preparation

Vol. 149: K. H. Hoffmann, W. Krabs (Eds.)
Optimal Control of Partial Differential Equations
Proceedings of the IFIP WG 7.2 International Conference, Irsee, April 9-12, 1990
VI, 245 pages. 1991

Vol. 150: L. C. G. J. M. Habets
Robust Stabilization in the Gap-topology
IX, 126 pages. 1991

Vol. 151: J. M. Skowronski, H. Flashner, R. S. Guttalu (Eds.)
Mechanics and Control
Proceedings of the 3rd Workshop on Control Mechanics in Honor of the 65th Birthday of George Leitmann
January 22-24, 1990, University of Southern California
IV, 497 pages. 1991

Vol. 152: J. D. Aplevich
Implicit Linear Systems
XI, 176 pages. 1991

Vol. 153: O. Hájek
Control Theory in the Plane
X, 272 pages. 1991

Vol. 154: A. Kurzhanski, I. Lasiecka (Eds.)
Modelling and Inverse Problems of Control for Distributed Parameter Systems
Proceedings of IFIP (W.G. 7.2)-IIASA Conference
Laxenburg, Austria, July 24-28, 1989
VII, 179 pages. 1991

Vol. 155: M. Bouvet, G. Bienvenu (Eds.)
High-Resolution Methods in Underwater Acoustics
V, 249 pages. 1991

Lecture Notes in Control and Information Sciences

Edited by M. Thoma and A. Wyner